PRENTICE-HALL

FOUNDATIONS OF DEVELOPMENTAL BIOLOGY SERIES

Clement L. Markert, Editor

Volumes published or in preparation:

FERTILIZATION *C. R. Austin*

CONTROL MECHANISMS IN PLANT DEVELOPMENT
Arthur W. Galston and Peter J. Davies

PRINCIPLES OF MAMMALIAN AGING
Robert R. Kohn

EMBRYONIC DIFFERENTIATION
H. E. Lehman

DEVELOPMENTAL GENETICS* *Clement L. Markert and Heinrich Ursprung*

CELL REPRODUCTION DURING DEVELOPMENT
David M. Prescott

CELLS INTO ORGANS: The Forces That Shape
the Embryo *J. P. Trinkaus*

PATTERNS IN PLANT DEVELOPMENT *T. A. Steeves and I. M. Sussex*

*Published jointly in Prentice-Hall's *Foundations of Modern Genetics Series*

Frontispiece   A plant of the Japanese morning glory (*Pharbitis nil*, variety Murasaki) induced to early flowering by short photoperiods (Chapter 1). In nature, these plants reach much greater heights before flowering, but since *Pharbitis* cotyledons are capable of sensing the length of day, even the first bud to mature can be floral. Normally, only one apical bud is produced; in this experiment that bud has been removed. This permits the normally inhibited cotyledonary buds to sprout (Chapters 3 and 5), and each has given rise to a beautiful violet flower. (Courtesy of Professors S. Imamura and A. Takimoto, Kyoto University.)

CONTROL MECHANISMS IN PLANT DEVELOPMENT

Arthur W. Galston

Yale University

Peter J. Davies

Cornell University

PRENTICE-HALL, INC. Englewood Cliffs, New Jersey

FOUNDATIONS OF DEVELOPMENTAL BIOLOGY SERIES

© 1970 by PRENTICE-HALL, Inc.
Englewood Cliffs, New Jersey.

Printed in the United States of America
C—13-171819-3
P—13-171801-0
Library of Congress Catalog Card Number: 78-127319

Current printing (last digit):
10   9   8   7   6   5   4   3   2   1

PRENTICE-HALL INTERNATIONAL, INC., London
PRENTICE-HALL OF AUSTRALIA, PTY. LTD., Sydney
PRENTICE-HALL OF CANADA, LTD., Toronto
PRENTICE-HALL OF INDIA PRIVATE LIMITED, New Delhi
PRENTICE-HALL OF JAPAN, INC., Tokyo

This book is affectionately dedicated to STERLING B. HENDRICKS and HARRY A. BORTHWICK of the United States Department of Agriculture, codiscoverers of phytochrome. Their collaboration provides a classic example of the benefits of a combined physicochemical and biological attack on a physiological problem. Their wisdom, humility and helpfulness have provided inspiration to an entire generation of experimental botanists.

Foundations of Developmental Biology

The development of organisms is so wondrous and yet so common that it has compelled man's attention and aroused his curiosity from earliest times. But developmental processes have proved to be complex and difficult to understand, and little progress was made for hundreds of years. By the beginning of this century, increasingly skillful experimentation began to accelerate the slow advance in our understanding of development. Most important in recent years has been the rapid progress in the related disciplines of biochemistry and genetics—progress that has now made possible an experimental attack on developmental problems at the molecular level. Many old and intractable problems are taking on a fresh appeal, and a tense expectancy pervades the biological community. Rapid advances are surely imminent.

New insights into the structure and function of cells are moving the principal problems of developmental biology into the center of scientific attention, and increasing numbers of biologists are focusing their research efforts on these problems. Moreover, new tools and experimental designs are now available to help in their solution.

At this critical stage of scientific development a fresh assessment is needed. This series of books in developmental biology is designed to provide essential background material and then to examine the frontier where significant advances are occurring or expected. Each book is written by a leading investigator actively concerned with the problems and concepts he discusses. Students at intermediate and advanced levels of preparation and investigators in other areas of biology should find these books informative, stimulating, and useful. Collectively, they present an authoritative and penetrating analysis of the major problems and concepts of developmental biology, together with a critical appraisal of the experimental tools and designs that make developmental biology so exciting and challenging today.

CLEMENT L. MARKERT

Preface

This brief volume is addressed to several audiences. First, it is our hope that advanced students of botany or plant physiology with some special interest in plant growth and development will find in the chapters that follow useful modern summaries of our understanding of the major regulatory mechanisms in plants. Much has been learned about this field within the last decade, especially at the phenomenological level, and it is useful to update published summaries of this knowledge for such students. Second, we hope that students of animal development will find in the subsequent pages an understanding of plant processes which will enlarge their overall view of morphogenesis in all organisms. Finally, and perhaps most importantly, we address this book to that new group of scientists called molecular biologists, most of whom have never thought about a plant as an object worthy of their scientific attention. Many of these people have been drawn from the ranks of physicists and chemists; they have ventured into the green pastures of biology, and have done much to revitalize and alter the path of development of that science. Through their efforts the most intimate details of the replication of viruses, of the synthesis of proteins and of the morphogenesis of organelles, viruses and even entire microbial cells have been elucidated. Thanks to them, we now understand much about mechanisms regulating the control of enzyme synthesis or repression, of the onset of gene activity in bacteria and insects and of the self-assembly of proteins and bacteriophages. Yet the plant world has remained relatively untouched by this revolution. Where physical chemists have turned their attention to the complex problems of the plant, remarkable new generalizations have emerged about the mechanisms through which light regulates plant activity. Some of these are detailed in the first chapter on *Phytochrome*. Where organic chemists have turned their attention to the isolation and identification of newly discovered regulatory substances, remarkable new insights into their molecular details have followed. Some of these results are detailed in the chapters on *Cytokinins, Gibberellins* and *Abscisic Acid,* which follow.

But this is just a beginning. The green plant, life's major link with the energy of the sun, is still a virtual unknown to molecular biologists. If any in that group happen to pick up this book (and we know that this is unlikely unless they receive special prodding), it is our hope that they will find described within these covers phenomena that may prove as diverting and rewarding to them as many of these problems have been to earlier students of plant development.

In the organization of this book we have deliberately adopted a radically new ordering of the material on plant development. This flows from two motivations. First, we believe that an occasional shaking up of the subject matter, standing the subject on its head in a way, forces people to look at all data in a new light. Second, since we discuss exclusively control mechanisms in plant development, the ordering of the subject matter is essentially immaterial, in the sense that the life cycle of a plant is indeed a continuum, and entry into the cycle may be made at any point. So, whereas in almost all books the discussion of plant reproduction and its control by phytochrome comes close to the end, we choose to begin with it, for it summarizes much beautiful recent work, and symbolizes what can be done by the combined attack of physicists, chemists and plant biologists on a common problem.

Within each chapter we have chosen to refer to a minimal number of references, for this book is not intended as an exhaustive review. Reviews are appearing in ever-increasing numbers, and anyone wishing an encyclopedic or thorough description of the recent literature in any one field should be referred to such publications as the *Annual Review of Plant Physiology* and the *Encyclopedia of Plant Physiology*. In citing a few key references and in providing minimal critical data to support the generalizations that we make in this book, we are hoping to paint a personal picture, i.e., our view of how the green plant controls its activities, integrates the functioning of part with part, and times its development with the cycle of the seasons in nature. If after reading this book the student is encouraged to read even one additional article cited in the literature, we shall feel that our labors have been worthwhile.

A word about prerequisite training for the use of this book is in order. Although it is our hope that any generally trained scientist and even the intelligent layman could pick up this book and read it with some interest and profit, clearly one's understanding of and ability to utilize this information constructively will be greatly enhanced by some general understanding of the green plant, especially its structure and morphogenesis. Any elementary exposition of this subject, such as the senior author's *Life of the Green Plant*, 2nd ed. (Prentice-Hall, 1964) will do. For the serious advanced student, we should point out that our omission of structural and ontogenetic facts has been deliberate. This volume is intended to accompany another volume in this series by T. A. Steeves and I. M. Sussex entitled *Patterns in Plant Development.* We hope the student will assimilate the contents of that book in order to understand this one more fully. Lastly, we have omitted details of the mechanisms of nucleic acid and protein synthesis and their control in the belief that advanced students of any phase of biology will have received instruction in these subjects elsewhere. Should this not be true, we would refer the reader to any modern textbook of biology or to the many excellent articles which have appeared over the last few years in the *Scientific American.*

New Haven, Connecticut ARTHUR W. GALSTON

PETER J. DAVIES

Contents

ONE

Phytochrome and Flowering

The green plant growing in the changing seasons of a temperate zone must synchronize its activities to suit the changing weather. Clearly, it would make no sense for a seed to start germinating immediately after it is produced on the mother plant in the waning days of the fall, for the rapid onset of winter would almost certainly freeze and kill the tender seedling produced. Similarly, a tree or shrub faced with the onset of cold weather must make some provision, before that weather arrives, for protecting the succulent growing points and tender leaves found in the buds. In deciduous trees this is accomplished by shedding the mature leaves, healing over the abscission zone by a corky layer, and resorting to special protective antiwinter mechanisms, such as cottony padding, waterproof protective scales, decreased succulence and higher osmotic concentration in the dormant winter buds. Finally, in initiating reproductive activity, it is clearly important for the plant to develop a reasonable size before it invests a large part of its photosynthetic product in the formation of fruits and seeds, yet it must permit sufficient time, following flowering, for adequate fruit and seed ripening before the start of the winter season.

To meet these challenges of timing, plants have evolved various mechanisms. One of the most prominent involves a pigment, named *phytochrome,* which exists in plants in at least two different forms, one active

1

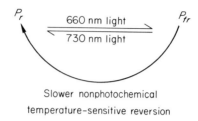

Fig. 1-1   Schematic representation of the transformations that occur between the red absorbing form of phytochrome (P_r) and the far-red absorbing form (P_{fr}).

and the other presumably inactive. Both forms absorb in the visible region of the spectrum, one in the red region at 660 nm (nanometers), the other in the "far-red" region at 730 nm. The form absorbing at 660 nm (P_r) is considered to be the inactive form. When P_r absorbs a quantum of light, it is transformed to the 730 nm absorbing form (P_{fr}). The latter, which is the active form of the pigment, may be transformed back to P_r, either by absorption of a quantum of light near 730 nm, or by thermal processes proceeding slowly in the dark (see Fig. 1-1). Since visible light emanating from the sun is a mixture of wavelengths causing the net transformation of P_r to P_{fr}, the alternation of light and dark periods in a normal day constitutes a system for the diurnal transformation in the state of phytochrome. The discovery and understanding of the wide significance of this simple chemical timing system constitute one of the more exciting stories of recent plant physiology.

Photoperiodism

It has long been appreciated that most plants, when grown at a particular latitude, flower at roughly the same date each year. We all have come to expect violets in the springtime, roses in the summer, and chrysanthemums in the fall. What is it that governs this behavior? Is the timing built into the seed? Is there an irreducible and unalterable number of days that must elapse before the vegetative plant, consisting of roots, stems and leaves, initiates floral primordia, or is there control by some environmental variable that has a given value at a given date? A simple experiment performed by Garner and Allard, immediately prior to 1920, answers this question (1). Various varieties of soybeans were planted at experimental stations located at different latitudes, or at successively later dates at a given latitude. Surprisingly, all plantings at a given latitude came into flower at the same time irrespective of planting date, (Fig. 1-2), while those planted farther and farther south initiated flowers sooner and sooner. Clearly, the flowering behavior of this plant is not regulated solely

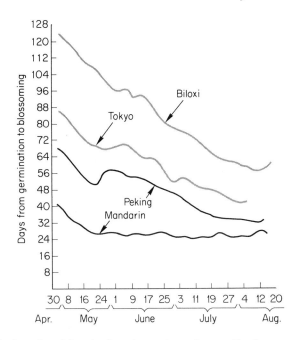

Fig. 1-2   The hastening of flowering in soybeans at one place resulting from progressively later planting during the growing season. Mandarin flowered at any day length and had only to achieve a minimum age before beginning reproduction. The other varieties flowered only as the day length decreased following midsummer, progressively shorter days being required by Peking, Tokyo and Biloxi, respectively. When Biloxi was sown before June 1, all the plants came into bloom within a few days at the beginning of September, showing an accurate measurement of day length. [From Garner and Allard (1).]

by internal events. If there is an external trigger, what is it? Again, Garner and Allard, building upon the leads furnished by other distinguished plant physiologists, such as Klebs in Germany, found that flowering in a new mutant of tobacco, referred to as Maryland Mammoth, is controlled by the day length. Very simply, if Maryland Mammoth plants were grown on long days (more than 14 hours of daylight) they did not flower. If grown on short days (fewer than 14 hours of daylight) they did flower (Fig. 1-3). They were thus referred to as *short-day plants*, and the length of the day that was barely short enough to promote the initiation of floral primordia was referred to as the *critical photoperiod*.

Later (2) Garner and Allard discovered that there are also *long-day plants*, i.e., plants that would not flower in short days but would flower in long days. Examples of this long-day type of plant are spinach and many cereals. (The distinction between long- and short-day plants is, however,

Fig. 1-3   Maryland Mammoth tobacco plants grown under short-day (left) and long-day (right) conditions. (From A. E. Murneek and R. O. Whyte. 1948. Vernalization and photoperiod. The Ronald Press Company, New York. Photo by Garner and Allard.)

not dependent on the absolute length of the critical photoperiod but on whether flowering is promoted by photoperiods longer or shorter than the critical photoperiod). Finally, they discovered a third group, the *day neutral plants*, exemplified by the tomato, in which flowering is relatively unaffected by the length of day, but is dependent on internal or structural features, such as the node number on the stem. Day length affects not only flowering behavior, but also many aspects of vegetative development, such as hairiness, production of anthocyanins, leaf fall, onset of dormancy, and the formation of underground food storage organs. These general responses of plants to the relative length of day and night are referred to collectively as *photoperiodism.*

What kind of mechanism regulates this behavior? One significant discovery by Hamner and Bonner in 1938 (3) was that in the cocklebur plant, which behaves as a short-day plant with a critical photoperiod of about 15 hours, the onset of reproduction is governed not by the length of the light period, but rather by the length of the dark period. This could be demonstrated most directly by placing the plants in artificially illuminated chambers. Whenever the dark period exceeded 9 hours, the plants flowered irrespective of the length of the light period. Conversely, whenever the dark period was less than 9 hours, the plants failed to flower, irrespective of the length of the light period. Clearly, a short-day plant is really a "long-night" plant, requiring some minimum length of uninterrupted darkness for flowering to occur.

If a dark period long enough to induce flowering in a short-day plant is interrupted near its midpoint by a flash of light of sufficient intensity, then its effect is negated and flowering does not occur (Fig. 1-4). The interruption of the dark period by light requires very little energy, certainly far less than would be required for photosynthesis. Experiments have shown that within broad limits, the plant obeys the reciprocity law or law of photochemical equivalence; that is, a given quantity of light energy (intensity × time) yields a given effect. A short flash of bright light is thus roughly equal to a long exposure to dim light in its inhibitory effect on flowering. This fact makes it possible to obtain an *action spectrum* for the process, and from this to deduce the nature of the absorbing pigment. The reasoning is as follows: Light to be effective must first be absorbed (the first law of photochemistry). Thus, when light energy is limiting to the overall rate of a photochemical process, the absorptive properties of the pigment will determine the relative effectiveness of the various regions of the spectrum in producing the effect. The careful application of selected energies of discrete wavelengths of light (Fig. 1-5) should permit one to make a comparison of the relative effect of the various portions of the spectrum on behavior, and from this to deduce the absorption spectrum of the pigment. In constructing such an action spectrum, one determines a dose-response curve for each wavelength, then calculates the amount of energy required to produce a standard effect, let us say 50% interruption of flowering compared with the control group. The reciprocal of such a figure would be a measure of the relative effectiveness of each wavelength.

Fig. 1-4 The effect of a light-flash interruption of the dark period on flowering in short-day and long-day plants.

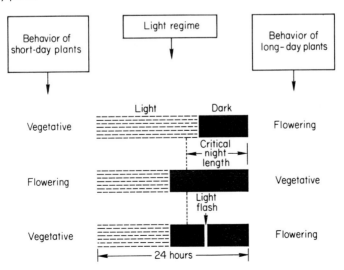

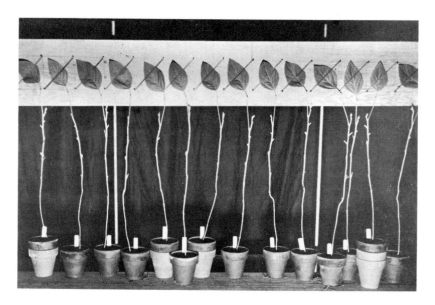

Fig. 1-5   The method of holding single leaves (these are soybean leaflets) in the image plane of a spectrograph for subsequent irradiation with various wavelengths of light. [From Hendricks and Borthwick (4).]

This may be refined to a per quantum basis by dividing the relative energy requirement by the average energy content per quantum at that wavelength.

Action spectra of this kind were produced for both long-day and short-day plants by Hendricks, Borthwick and collaborators at the United States Department of Agriculture at Beltsville, Maryland (4). A representative spectrum is plotted in Fig. 1-6. It should be noted that the maximum effect is produced in the red region of the spectrum, between 620 and 680 nm, that there is a diminishing effect beyond 700 nm, little effect in the yellow-green portion of the spectrum near 500 nm, and some slight effect in the blue region near 400 nm.

The discovery of phytochrome

What kind of pigment has an absorption spectrum to match this action spectrum? Clearly, this could not be chlorophyll, which in addition to a red peak also has a large absorption peak called the *Soret* peak in the vicinity of 430 nm (Fig. 1-7). The closest guess one might make is that something resembling a bile pigment, or phycobilin, might be involved. Such pigments contain four pyrrole groups, as do chlorophyll and heme, but instead of being joined cyclically, they are opened out into a straight chain. Such molecules have the red peak characteristic of chlorophyll but

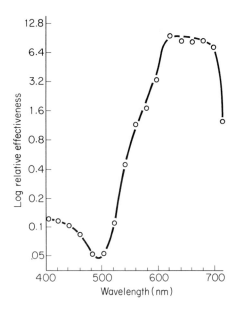

Fig. 1-6   A generalized action spectrum for photoperiodism. (Calculated from data of Parker, Borthwick, Hendricks and Went. 1949. Amer. J. Bot. **36**:194–204.)

lack the Soret peak. An example of such a compound is the phycocyanin of a blue-green alga consisting of an open chain tetrapyrrole attached to a protein (Fig. 1-8). Many early attempts to extract such a pigment from higher green plants resulted in failure.

In view of the fact that in green plants chlorophyll predominates and completely swamps out the absorption properties of the hypothetical receptor pigment, Hendricks and Borthwick turned to other systems, devoid of chlorophyll. They found that the light control of morphogenesis in various etiolated plants such as the pea and albino barley has an action spectrum almost identical with that for the control of flowering in long-day and short-day plants. This strengthened the view that the action spectrum represented a fairly close approximation of the absorption spectrum of the effective pigment, unaltered by interfering pigments such as chlorophyll and the carotenoids. Yet even in etiolated and albino plants, early attempts to extract the pigment failed.

There are many examples in the history of science in which unexpected findings clear up problems that have long puzzled investigators. The final elucidation of the nature of phytochrome came from an old observation on an apparently unrelated problem. A generation earlier, at the Smithsonian Institution, Flint and McAlister had studied the effect of light on the germination of various varieties of lettuce seeds (5). Such seeds, when

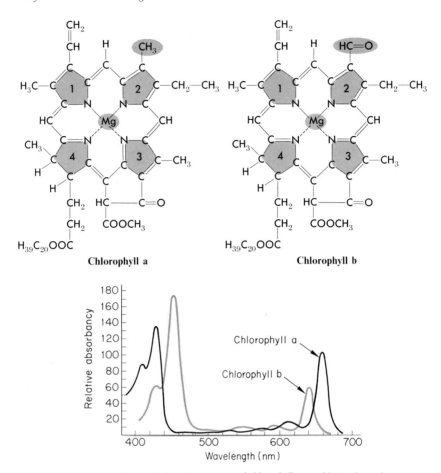

Fig. 1-7   The formulae and absorption spectra of chlorophylls a and b in ether solution.

placed in darkness, germinate very poorly, sometimes not at all. If, however, they are given some light after they have imbibed water, germination proceeds rapidly and can be measured within 24 to 48 hours. An action spectrum for the promotion of germination shows a large peak in the red region of the spectrum, just where Hendricks and Borthwick later found maximal photoeffects on flowering. Another feature of significance is that in the region beyond 700 nm Flint and McAlister found a marked *inhibitory* effect on germination. This could best be demonstrated by bringing the seeds up to about 50% germination with weak white light, after which both the promotive effects of red and the inhibitory effects of what came to be called far-red light were manifest (Fig. 1-9).

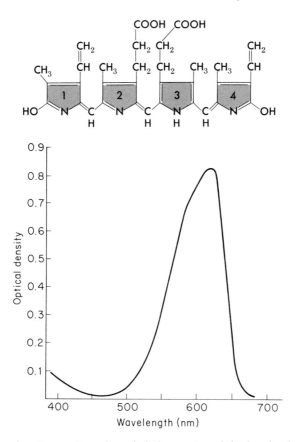

Fig. 1-8 The absorption spectrum of an algal phycocyanin and the formula of a phycobilin (biliverdin), which is the chromophore in a phycocyanin.

Borthwick and Hendricks seized upon this observation as a guide to further experiments. What would happen, they asked, if seeds were exposed first to red and then to far-red light, or first to far-red and then to red light? The answer to this question is shown in Table 1-1 (6). It is clear that lettuce seeds may be promoted to germinate by red light, and that the effect of red may be negated by subsequent far-red light; the reversing effect of far-red may in turn be negated by red light. It is as if there were a two-way switch capable of being pressed any number of times without marked effect on its operation; the plant responds only to the last press of the switch. Not only does this system work in lettuce seed germination, but it is also effective in the control of flowering (Fig. 1-10). From these data, Hendricks and Borthwick proposed that the pigment which they

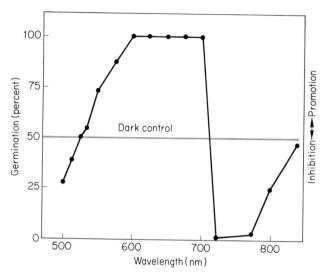

Fig. 1-9　Seeds of light-sensitive lettuce induced to 50% germination by limited red light can be further stimulated to germinate by red light (600 to 690 nm) or inhibited by far-red light (720 to 780 nm). [From Flint and McAlister (5).]

were seeking exists in two mutually photoreversible forms. Form 1 has an absorption peak in the red and very little absorption in far-red light; form 2, on the other hand, absorbs heavily in far-red and less heavily in red light. Both forms have a small peak in blue light. The red absorbing form of the pigment, called P_r, is transformed to P_{fr}, the far-red absorbing form of the pigment, by a quantum of 660 nm light. The reverse transformation is produced by 730 nm light.

TABLE 1-1

THE GERMINATION OF LETTUCE SEED AFTER EXPOSURE
TO RED (R) AND FAR-RED (FR) LIGHT IN SEQUENCE

Irradiation	*Germination* (%)
None (dark control)	8
R	98
R + FR	54
R + FR + R	100
R + FR + R + FR	43
R + FR + R + FR + R	99
R + FR + R + FR + R +FR	54
R + FR + R + FR + R + FR + R	98

From Borthwick, et al. (6).

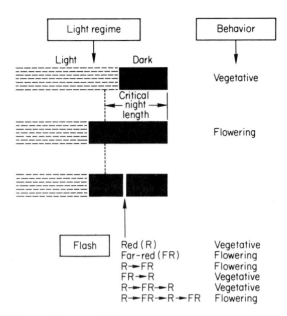

Fig. 1-10 The reversible control of flowering by red and far-red light in short-day plant.

This hypothesis, besides explaining the data for lettuce seed germination and the control of flowering and morphogenesis by light, also permitted the construction of a physical assay for the pigment. A light-measuring device was constructed which determined the change in absorbancy at the two relevant wavelengths, 660 and 730 nm. By special arrangements of circuitry and placement of the photocell, small absorbancy changes could be picked up against a background of high total optical absorption. With this instrument it could be shown that in etiolated tissue light absorption decreased at 660 nm and increased at 730 nm when the plant was irradiated with 660 nm light, whereas irradiating the tissue with 730 nm light resulted in a decrease in optical density at this wavelength and a concomitant increase at 660 nm. The hypothesis had been brilliantly confirmed by the experimental data obtained in vivo.

Now it was only a question of time until suitable extracts could be made which would mimic in vitro the data obtained in vivo. At the Beltsville laboratories, a team of Borthwick and Hendricks' colleagues, including Siegelman, Butler and Norris, succeeded in extracting and partially purifying a pigment possessing many of the properties of the in vivo system (Fig. 1-11) (7). This pigment was named *phytochrome.* The purest preparations of phytochrome have shown it to be a protein with a molecular weight of about 60,000, containing one prosthetic group of the open-chain tetrapyrrole type. There is some evidence, both from in vivo and in vitro

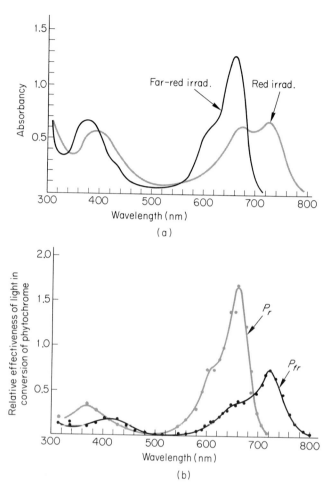

Fig. 1-11   The (a) absorption and (b) action spectra of solutions of phytochrome. The absorption spectra were taken following irradiation with either red or far-red light to convert the phytochrome to the form that absorbs in the far-red (730 nm) or red (660 nm), respectively. Note that the spectra of the two forms overlap. Phytochrome also absorbs in the blue region, producing effects intermediate between those of red and far-red light. The action spectra show the effectiveness of different wavelengths of light in converting phytochrome to the alternative form. Although light at 660 nm is predominantly effective in changing P_r, it is also effective in converting P_{fr}. This is due to the overlapping spectra noted above. Thus when P_r is irradiated with red it is converted to P_{fr}, but some of the P_{fr} is also reconverted into P_r, so that the absorption spectrum of red irradiated phytochrome shows that we have a mixture of predominantly P_{fr} and some P_r. (From Siegelman and Butler. 1965. Ann. Rev. Plant Physiol. **16**:383–392.)

work, that phytochrome may exist in several different forms in the cell, differing in molecular weight, kinetics of conversion and stability.

An amino acid and sulfhydryl analysis of the protein has been made (8), indicating that phytochrome closely resembles phycocyanin in basic structure. It appears that the P_{fr} form is much less stable than the P_r form, in that it is much more easily denatured by high concentrations of urea, and is more readily attacked by such agents as p-chloromercuribenzoate, known to attach to SH groups. On the other hand, the P_r form is more susceptible to attack by aldehydes such as formaldehyde and glutaraldehyde. High-energy flash photolysis experiments with sophisticated equipment have also revealed the existence of a number of transient intermediates between P_r and P_{fr}. It appears that P_r and P_{fr} undergo transformations through different intermediates when excited by photon capture.

Although the properties of phytochrome, as studied in vivo and in vitro, satisfactorily explain the action of light on the control of growth and form in many plants, there are some yet unresolved paradoxes that must be mentioned. For example, in some plants that are known to respond to red and far-red light, no phytochrome has yet been detected. This is generally considered by most experts in the field to be due to some trivial technical problem that will ultimately be solved. More serious is the situation in which the effect of red light on plant growth is completely saturated by energies so low that no detectable phytochrome conversion can be shown to occur. The same paradox has been found in reverse in plant systems in which the complete conversion of phytochrome from P_r to P_{fr} fails to remove sensitivity to red light. These paradoxes have led to the hypothesis that not all the phytochrome in a plant is active in morphogenesis. Rather, only a small fraction of the total, perhaps attached at a particular locus or to a particular protein in the cell, is effective. The rest, unattached, can show the spectral transformation characteristic of phytochrome, yet be ineffective in controlling growth and form. Recently the discovery of two or more kinetically distinguishable populations of photoreversible phytochrome, possibly differing in molecular size, has added evidence to the above ideas. Should these be shown to differ in physiological activity, resolution of the paradoxes would not be difficult, but whatever the answer to these paradoxes, they do not significantly alter the validity of the phytochrome theory of the control of plant form.

Mode of action of phytochrome

What does phytochrome do to control plant behavior? In the first place, we must recall that phytochrome controls an amazing variety of processes, ranging from seed germination and the control of leaf and stem growth to the development of plastids and the control of bolting and flowering in a great many plants. Certainly it is simplest to believe that if phytochrome is basically the same in all plants (as we have every reason to expect it is), it must have a single primary action common to all the morphogenetic processes controlled by phytochrome. At first, it was thought that such a process might be represented by derepression of a genetic locus. This was hypothesized because in certain phytochrome-controlled systems, the action of light on morphogenesis can be prevented by feeding to the plant known inhibitors of the synthesis of RNA and proteins (9). Although it is certainly true that inhibitors like actinomycin D, puromycin, chloramphenicol and analogs of nucleotides and amino acids do interfere with phytochrome-controlled morphogenesis, it now seems untenable to attribute this to *direct* interference with the phytochrome system. Certainly if the plant is going to make flowers, it is going to have to make new cells, and if new cells are made, new RNA and protein species must be constructed. Interference with this process will naturally interfere with the end result of phytochrome action, but recent evidence suggests that phytochrome acts much more rapidly in quite another way.

As an example of a rapid response controlled by phytochrome, let us examine the nyctinastic or sleep movements of such leguminous plants as the well-known sensitive plant, *Mimosa pudica,* and its nonthigmonastic relative, *Albizzia.* In both of these genera, the pinnules of the doubly compound leaves fold together in darkness, but are open and separate in the light. These movements are known to be controlled by the turgor of cells of the pulvinules, which attach each leaflet to a rachilla. When such cells are turgid, the leaves are open; when they are flaccid, the leaves are closed. It has now been shown that this nyctinastic movement, which can be observed within ten minutes, is under the control of phytochrome. Thus, giving a plant red light before transferring it to darkness permits rapid closure, whereas giving the same plant far-red prevents closure (Fig. 1-12) (10). It thus appears that phytochrome must be in the P_{fr} form if closure is to occur.

Several other rapid movements in response to phytochrome are known. For example, phytochrome-controlled plastid orientation changes in the alga *Mougeotia* (Fig. 1-13) are evident in less than 10 minutes and are fully displayed by 30 minutes following irradiation with red light. The response to red light can be fully reversed by far-red light only during the first minute after red irradiation. The potentiation of leaf unrolling in etiolated grasses and of flowering in *Pharbitis* and *Kalanchoë* occurs within about one minute, though the result may take several days to appear. If lettuce seeds are germinated in the presence of gibberellic acid, conversion of phyto-chrome to the P_{fr} form for only 5 minutes greatly promotes the process, though the germination response cannot be seen until far later. Although

Fig. 1-12 Pinnae of *Mimosa pudica* 30 minutes after placing in darkness following high-intensity light terminated with a succession of 2-minute exposures to red (R) and far-red (FR) irradiation. The pinnae remained open if exposure to far-red was last (top row) and closed if red irradiation was last (bottom row). [From Fondéville, et al. (10).]

these very rapid potentiation effects are not conclusive evidence against phytochrome action via a relatively slow process like protein synthesis, it seems more likely that the primary action of phytochrome concerns a more rapid process, in which ribonucleic acid and protein synthesis are not involved. This conclusion is further strengthened by the fact that the phytochrome-mediated closure of *Albizzia* leaves is in no way affected by actinomycin D.

One reasonable hypothesis to explain these phenomena is that phytochrome is localized in the membrane, and that its transformations affect the state of that membrane's permeability; all subsequent morphogenetic transformations would be secondary derivatives of this primary effect. The way in which the membrane permeability is changed is uncertain, but one of the results appears to be the induction of a localized electrochemical gradient that manifests itself as a bioelectric potential. This was discovered

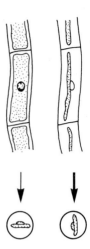

Fig. 1-13   Diagram of the alga Mougeotia (longitudinal section above, transverse section below). Weak red light causes the chloroplast to orient face-on towards the light. Far-red light or strong white light causes the chloroplast to turn side-on to the light. (From Haupt. 1963. Ber. der Deutsch, Botan. Gesellschaft **76**:313–322.)

from the demonstration that an excised mung bean root tip developed the ability to adhere to a negatively charged glass surface when irradiated for 4 minutes with red light, and this was reversed by far-red light. Examination of the electrical potentials on the root tip showed that red light caused the induction of a small positive charge and that this was reversed by far-red light (Fig. 1-14) (11).

Localization of phytochrome

If phytochrome acts on such a system as membrane permeability, its location within the cell must be of considerable significance. Studies on the localization of phytochrome are still in their infancy, yet some significant information is at hand. One method used to investigate phytochrome localization has involved the use of a microbeam of light directed at specific parts of the cell to detect light absorption characteristic of phytochrome and the expected red, far-red photoreversible absorbancy changes. With this technique, phytochrome has been detected in the vicinity of the nuclear membrane of cells from oat coleoptiles and pea stems (12). The amount present is, as would be expected, extremely small, but reversible changes in absorption spectrum were brought about by repeated alternation of the irradiation from 660 nm to 730 nm (Fig. 1-15). No phytochrome could be detected elsewhere in the cell with this technique. This does not mean that phytochrome is absent from other regions

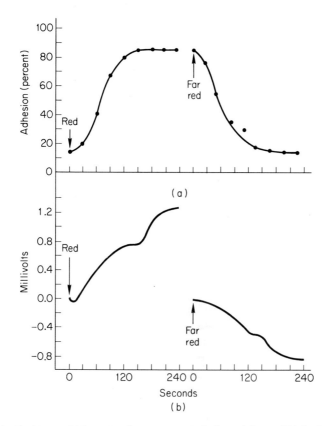

Fig. 1-14   The kinetics of (a) root tip adhesion to negatively charged glass and (b) the development of electrical potentials in the root tips following irradiation with red or far-red light. [From Jaffe (11).]

of the cell; it may be either masked or present in too small an amount to be measured.

Phytochrome has also been localized in the large cell of the alga *Mougeotia*, where red light induces a turning of the chloroplast (13). To achieve this response, it is not essential to irradiate the chloroplast itself. By focusing the microbeam on different parts of the cell, Haupt deduced that the locus of the reception for this stimulus is the cell membrane. If only a small portion of the cell membrane is irradiated, only a small portion of the chloroplast nearest the membrane responds (Fig. 1-16). This shows that the stimulus in this cell produces only a localized effect and implies that the nucleus has no direct role in the response. By the use of polarized light with various orientations it has been shown that phytochrome has a fixed orientation within the cell membrane, and it has been

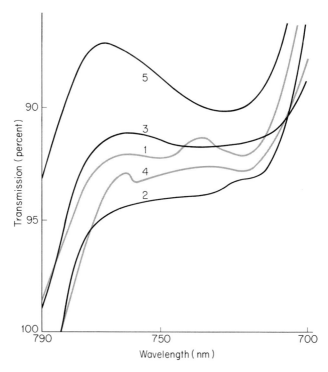

Fig. 1-15   Far-red absorption spectra of the nucleus from etiolated pea epicotyl following alternating 2-minute light treatment. 1: original trace. 2: FR; 3: FR, R; 4: FR, R, FR; 5: FR, R, FR, R. Note that far-red light last lowers whereas red light last increases the absorption of far-red light. [From Galston (12).]

suggested that irradiation may change the orientation of the phytochrome molecules according to the wavelength of the light applied.

High-energy reactions

Besides the paradoxes mentioned earlier, it has become clear that all the morphogenetic results of red and far-red irradiation cannot be explained upon the basis of the simple phytochrome red/far-red reversibility. Many changes in morphogenesis respond positively to both prolonged high-intensity red and far-red light, and have thus been termed *high-energy reactions.* They have an action spectrum with peaks in the region of 700 to 720 nm and also in the blue region of the spectrum (Fig. 1-17). The high-energy reaction is clearly of importance in plant development in that it is active in phenomena such as inhibition of hypocotyl elongation and promotion of anthocyanin formation in seedlings, and flowering in

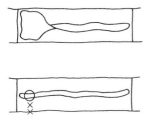

Fig. 1-16   The result (above) of irradiating Mougeotia with a microbeam (3 μ diameter) (below) shows that the localization of the phytochrome receptor of the light is not in the chloroplast (as beams outside the chloroplast also produce movement) and not in the nucleus, which is in the center of the cell. One possible localization is the cell membrane. (From Bock and Haupt. 1961. Planta **57**:518–530.)

long-day plants. What is the pigment system controlling these changes? The properties of the system do not match those of the reversible phytochrome responses, though it has been suggested that phytochrome may be involved in other ways. An explanation has been advanced that constant illumination with far-red light leaves a small percentage of the phytochrome in a destruction-resistant P_{fr} form and that reactions in this category require higher activation energies than other phytochrome responses. Alternatively, the action spectrum of the responses may be due to light absorption by intermediates of phytochrome between the P_r and

Fig. 1-17   The action spectrum of the high-energy reaction in mustard seedlings. (From Mohr. 1962. Ann. Rev. Plant Physiol. **13**:465–488.)

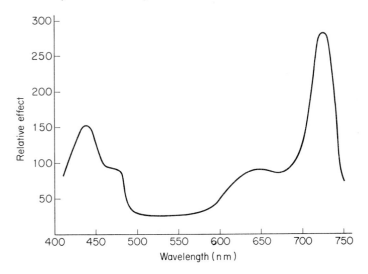

P_{fr} forms. In high-intensity light, such as sunlight, it appears that approximately 30% of the phytochrome may be in the form of intermediates because of the rapid conversion of P_r to P_{fr}, and vice versa. This would result in an absorption spectrum with an intermediate peak, and it is possible that the intermediates themselves exert a positive effect on morphogenesis. Neither of these explanations, however, considers the absorption peak in the blue, and it may be that the high-energy responses are mediated by some other, as yet undiscovered, pigment.

Localized effects of
phytochrome conversion

As we have already seen, the conversion of phytochrome from P_r to P_{fr} can influence almost every aspect of plant development from seed germination to the formation of reproductive organs. While all of these phenomena may, as we have suggested, be controlled ultimately by a single physiological parameter such as membrane permeability, there are important differences among the various responses, which require some discussion. Probably the most important difference has to do with localization versus propagation of the effect.

Let us consider a lettuce seed, for example. Germination is usually assayed as the protrusion of the radicle outside the seed coat. In fact, if microbeams of red light are caused to impinge on a lettuce seed, it can be shown that illumination of the radicle end is most effective in the promotion of germination. From this, it may probably be concluded that phytochrome conversion in the radicle itself is a prerequisite for germination. By implication one is led to the conclusion that conversion of phytochrome in, for example, the cotyledon would not have any effect on the phytochrome of the radicle and hence on the subsequent germination process.

The same sort of localization seems to apply to the control of morphogenesis in seedlings. In the young etiolated pea plant, red light inhibits the growth of internodes, promotes the growth of the plumular leaves, and opens the apical hook. Experiments with microbeams of red light have shown clearly that only that part which is irradiated responds, so that the enhanced growth of the plumule is not a consequence of the decreased growth of the epicotyl. Furthermore, if one excises pieces of epicotyl and of plumule, one can show the same effects of light on the excised parts as are apparent in the intact organism. Thus, in the pea seedling, as in the lettuce seed, it appears that light must hit the cell whose growth pattern is to be altered; no transmitted effects are noted. In the light control of

Fig. 1-18   Part of a leaf of *Albizzia* after transfer from light to darkness. The pulvini at the base of second, fourth and sixth leaflets from the top were exposed to 2 minutes of far-red light immediately before the dark period. It is clear that far-red light has prevented the closure of only the leaflets whose pulvini were irradiated. [From Koukkari and Hillman (14).]

nyctinasty of *Albizzia* leaves, the receptor of the light stimulus is the pulvinus itself, which also shows the response (14). There is no translocation of the stimulus, so that with localized light stimulation it is possible to stimulate individual leaflets to close while others remain open (Fig. 1-18).

Propagated effects of
phytochrome conversion

The control of flowering by phytochrome conversion introduces a new set of circumstances, for here the effects of phytochrome conversion are propagated over long distances, and are perpetuated for long periods in time. Let us consider the short-day plant *Xanthium* (cocklebur), which will flower if the uninterrupted dark period exceeds 9 hours. Suppose that we put one part of the plant on short day by enclosing it in some light-tight container, while the rest of the plant is exposed to long day. If there were no transmitted effects, one would expect only the portion exposed to short day to flower. However, all buds may be induced to flower, even those on the portion of the plant exposed to long days. Clearly, some influence has moved from the portion exposed to short day to the portion exposed to long day.

This experiment can be refined to the point where only a single leaf enclosed in the opaque container is receiving a short day, while the rest of the plant receives a long day. Here again buds all over the plant can be induced to flower (Fig. 1-19). This shows clearly that the leaf is the locus of photoperiodic perception and that it influences the bud some distance away. The influence is almost certainly due to a chemical substance synthesized in small quantities by the photoinduced leaf. This substance, the hypothetical flowering hormone of plants, has been named *florigen*. Florigen has never been isolated and chemically identified, but its presence can be shown in a variety of ways. For example, if plant A is induced to flower and plant B is not, flowering may be caused in B by grafting it to A (Fig. 1-20) (3). Even a single leaf from the donor grafted on to the receptor may suffice. Through the use of steam girdling techniques, which kill cells and also block florigen transfer, it is deduced that the flow of hormones from donor to receptor occurs in the vascular system,

Fig. 1-19   In many species, as in the cocklebur, short-day treatment of a single leaf results in flowering of the plant. [Adapted from Hamner and Bonner (3).]

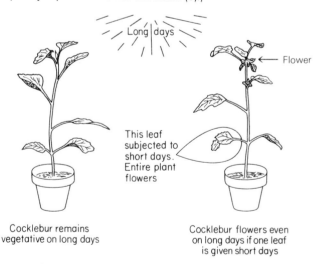

Cocklebur remains
vegetative on long days

Cocklebur flowers even
on long days if one leaf
is given short days

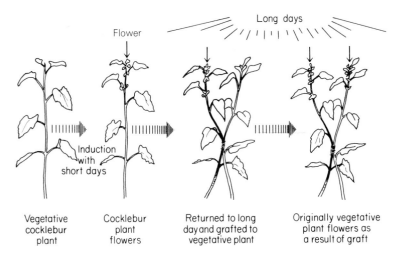

Fig. 1-20 The flowering stimulus may be transmitted from plant to plant across a graft union. [Adapted from Hamner and Bonner (3).]

presumably in the living cells of the phloem, at about the rate of transport of bulk organic substances in plants.

There have been occasional disputed reports that application of leaf extracts to receptor plants has resulted in flowering. One recent report (15) of the initiation of flowering in *Lemna* (duckweed) by extracts of flowering *Xanthium* but not by extracts of vegetative *Xanthium* indicates that we may be coming nearer to an understanding of the nature of florigen. The extracts also produced flower initiation in *Xanthium,* but only in the presence of supplemental gibberellic acid. The application of gibberellin to long-day plants under short days generally causes a stem elongation, or bolting, frequently followed by flowering (Fig. 1-21), though not in all cases. It is, therefore, possible that gibberellin may be the flowering hormone in certain long-day plants, but it has no effect on short-day plants. There is now evidence that abscisic acid can promote flowering in some short-day plants. Since both of these substances are made of isoprenoid units, some people have speculated that phytochrome conversion controls some aspect of isoprenoid biosynthesis. Florigen would then be a different compound depending upon the nature of plant's response to day length, and we do know that gibberellins are synthesized in response to long days and abscisic acid to decreasing day length (see Chapters 4 and 5). The effects of these two hormones are, however, far from universal, so the exact nature of florigen still remains a mystery.

Fig. 1-21   The long-day-requiring plant *Samolus parviflora* can be induced to flower under noninductive short days (9 hours) by applications of gibberellin. From left to right, plants were given 0, 2, 5, 10, 20 or 50 μg/plant each day. (From Lang. 1957. Proc. Nat. Acad. Sci. **43:**709–717.)

Phytochrome and florigen

How can one reconcile transmission of the response to phytochrome changes in some cases with lack of transmission in others? If phytochrome changes some very basic physical property of the cell such as membrane permeability, this not only could influence water relations but might cause the activation of enzyme systems, resulting in the production of new substances, including hormones. In leaves, red light has been shown to induce gibberellin formation within 30 minutes (16). However, exposure of lettuce seeds to short periods (5 to 30 minutes) of P_{fr} with added gibberellic acid produced synergistic and not additive effects on germination (Fig. 1-22). This would appear to eliminate gibberellic acid formation as the basic action of red light in seeds (17); at a minimum, it shows that phytochrome is doing something in addition to promoting gibberellin formation. An increase in membrane permeability, which might be the route through which increased gibberellins are synthesized, would also allow increased penetration and thus increased action of the applied gibberellic acid. The most we can say is that phytochrome controls several kinds of phenomena that may or may not be due to the same primary action.

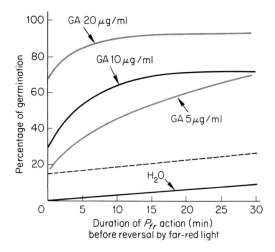

Fig. 1-22 The action of P_{fr} in enhancing germination of lettuce seeds is synergistic in the presence of added gibberellin. The dotted line shows the anticipated germination if the effects of gibberellin and P_{fr} should be additive. [From Bewley, et al. 1967 (17).]

Interaction of phytochrome with other controlling factors

We have mentioned earlier that not all plants show photoperiodic control of flowering. Such day-neutral plants do not lack phytochrome, since one can demonstrate red, far-red reversible control of some aspects of their vegetative morphogenesis. This tells us that phytochrome conversion is not always linked to the system that controls reproduction. Even some photoperiodically sensitive plants fail to respond to photoperiod at particular times in their life history. For example, some plants must pass from a juvenile phase to a phase designated as "ripeness to flower" in order for photoperiod to be effective; in such plants, the green expanded cotyledons or even first leaves are not sensitive to photoperiod, whereas later leaves are. Failure to respond is once again not due to the absence of phytochrome, since such seedlings show the usual morphogenetic responses to red light. Rather, some other aspect of the response apparatus would appear to be undeveloped. The same sort of conclusion is derived from the fact that additional requirements for flowering may become less rigid as plants age (Fig. 1-23).

Floral induction

We have mentioned earlier that the response to the length of uninterrupted night behaves like an "all or nothing" phenomenon. For example, in a short-day plant flowering is not initiated *at all* until the

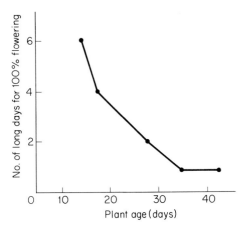

Fig. 1-23   The long-day requirement for flower induction in *Lolium temulentum* declines as the plants increase in age. (Data of Evans. 1960. Aust. J. Biol. Sci. **13**:123–131.)

critical length of dark period is attained. However, there are certain quantitative aspects of floral induction that need further discussion. If we once again consider *Xanthium,* the cocklebur plant, we note that there are various anatomical stages involved as the plants progress from vegetation to complete flowering (Fig. 1-24). If these stages are arranged in order, they constitute a quantitative expression of the degree to which flowering has progressed. Using this tool, many investigators have been able to show that increasing lengths of dark period beyond the critical lead to steadily higher scores in the flowering assay. Clearly, then, some reaction is proceeding progressively as the length of the critical dark period is exceeded.

How many cycles are required to cause flowering? In the cocklebur it appears that only one such period suffices, at least for the induction of minimal reproduction, although more than one photoinductive cycle can give a greater total effect (Fig. 1-25). With some short-day plants, such as soybeans, approximately four successive photoinductive cycles are required, while other species may require even more (Fig. 1-26). If these are not given sequentially, then strict additivity is not observed, and more than the number of successively effective cycles will be required.

What happens following floral induction to cause an indefinite prolongation of flowering, even when the plant is returned to photoperiods unfavorable for flowering? Part of the answer undoubtedly has to do with the fact that once basic anatomical transformations occur in the meristem, there is no way to reverse them, and that after they have occurred the normal processes of plant growth contribute to the formation of floral rather than vegetative primordia. But there is more to it than this, for

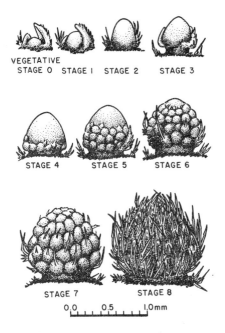

VEGETATIVE
STAGE 0   STAGE 1   STAGE 2   STAGE 3

STAGE 4   STAGE 5   STAGE 6

STAGE 7   STAGE 8
0.0   0.5   1.0 mm

Fig. 1-24   Stages of development in the stem apex of *Xanthium* from vegetative growth to a complete inflorescence primordium. (From Salisbury. 1955. Plant Physiol. **30**:327–334.)

Fig. 1-25   If the floral stages (Fig. 1-24) are used as a measure of the influence of light treatments on flowering in *Xanthium*, continuous short days are seen to cause a more rapid floral development than occurs after only two short days. (From Salisbury. 1955. Plant Physiol. **30**:327–334.)

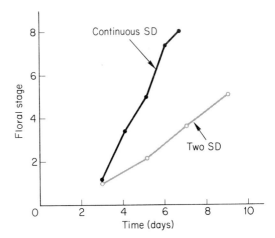

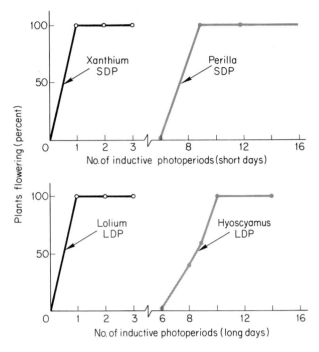

Fig. 1-26   Some plants such as *Xanthium* and *Lolium* require only one inductive cycle for flowering, whereas others, such as *Perilla* and *Hyoscyamus*, flower only after several such cycles. Expressed as a percentage of plants flowering, the response to photoperiodic induction acts nearly like an all-or-none response, though some species that require several cycles may initially show a limited quantitative response followed within a few further cycles by complete flowering. (Data from Evans. 1960. Aust. J. Biol. Sci. **13**:123–131. Lang and Melchers. 1943. Planta **33**:653–702. Schwabe. 1959. J. Exp. Bot. **10**:317–329.)

plants that have been induced to flower may now serve as sources of the floral stimulus and may, in fact, continue to act as donors in grafting experiments far beyond the time of their initial induction. Thus it appears that transition from vegetation to reproduction involves some kind of metabolic transition, leading to the prolonged and perhaps indefinite production of the floral-inducing stimulus. This may result from the perpetuated induction of an enzyme system, or it may represent the unmasking of a self-perpetuating organelle, analogous to the episomes of microbial cells. Let us recall, for example, that some bacterial viruses (prophages), replicate within the cell but are unnoticed because they do not replicate complete virus particles. If such cells are irradiated with ultraviolet light, the prophages are converted into active phages, which then form complete viruses and destroy the host cell. So it may be that the floral stimulus represents some kind of latent self-replicating entity that is transformed into the active form by appropriate photoperiod. In this context, the initiation of flowering could be compared to catching a disease, which is then propagated from one individual to another, but occasionally is terminated by the spontaneous recovery of the host. Thus in some plants, such as the short-day *Perilla*, minimal induction leads to flowering for a time, but this is followed by a gradual return to vegetative growth, which terminates the induction.

Endogenous rhythms and the
effects of phytochrome

How does the plant measure time with regard to the length of the night necessary for the induction of flowering? In addition to its conversion to P_r by far-red light, P_{fr} also reverts slowly to P_r in the dark. This may be direct or may involve destruction of the P_{fr} and resynthesis of the P_r. By destruction we mean simply a decrease in the total phytochrome as measured by differential two-wavelength spectrophotometry. Actually, phytochrome might not really be destroyed but rather changed in some way, such as separation of the protein and chromophore groups, or a binding to some particulate matter that renders it nonphotoreversible and therefore not detectable. In like manner, the resynthesis of P_r may be simply a reactivation of the previously inactive molecule. It was originally suggested that this reversion of P_{fr} to P_r was the timing mechanism in photoperiodism. Recent experience indicates that this is, however, not the controlling mechanism, and what is important is an interaction of the state of the phytochrome with endogenous rhythms in the plant (18). Before discussing these interactions we should examine the nature of endogenous rhythms.

Many experiments have suggested that all higher plants contain in their cells endogenously oscillating systems or *biological clocks* that govern rhythmic alterations in behavior. Some of these rhythms, being approximately 24 hours long, are referred to as *circadian*. One example is the sleep movement of leaves, studied most extensively by the German plant physiologist, Erwin Bünning. The basic experiment is as follows: if a bean plant that has been grown in a greenhouse is transferred to a dark room, its leaves will continue their rise and fall (horizontal in the light, vertical in the dark) in an approximately 24-hour rhythm. This movement may easily be visualized by tying a silk thread onto the leaf tip, bringing it up over a pulley, and causing a record of leaf movements to be made on a rotating drum (Fig. 1-27). If the dark period during which the leaves would normally be "asleep" is punctuated by a flash of light, then the rhythm is interrupted and the leaves tend to return to the horizontal position (Fig. 1-28). Administration of a flash of light during that time of the cycle

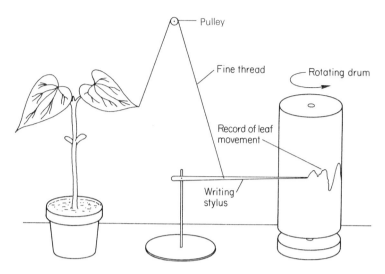

Fig. 1-27   A recording of endogenous movements of bean leaves.

when the leaves would normally be "awake" does not have any such resetting effect. Thus it appears that the effect of a light flash depends on the phase of the endogenous rhythm in which it is administered.

Bünning later found that, in many varieties of soybeans, there was a marked relationship between sleep movements of their primary leaves and their photoperiodic behavior. Those plants that had very definite sleep

Fig. 1-28   Circadian movements of a leaf of a bean plant (*Phaseolus vitellinus*) shows a rising in the day and a falling at night with the movements starting before the actual change in light conditions. A 24-hour light period (a) caused a resetting of the rhythm (b) so that the new rhythm is then followed. (From Bünning. 1959. Encyclopedia of Plant Physiol. **17** (1):579–656.)

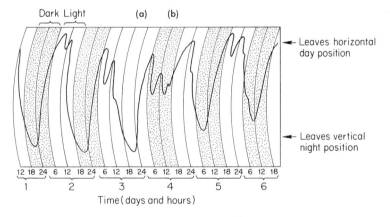

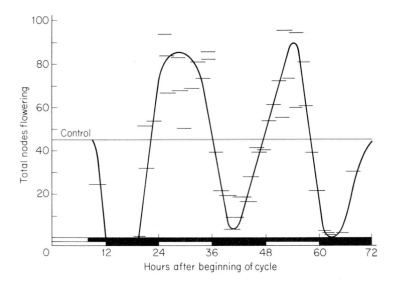

Fig. 1-29 Responses of Biloxi soybean plants to 4-hour light interruptions (indicated by the lines), given at various times during a 72-hour cycle (8 hours light and 64 hours dark, as indicated by the upper bar at the bottom of the graph). The lower bar indicates the daily cycle of light and darkness under normal conditions. The control (gray line) received 8 hours light and 64 hours dark. (After data of Hamner, from Salisbury. 1965. Endeavour **24**:74–80.)

movements tended to be short-day in photoperiodic behavior. He thus hypothesized that the requirement for a specific photoperiod for the onset of flowering was in some way directly related to the state of a plant's endogenous rhythms. This theory has been tested by several investigators, especially K. C. Hamner. Hamner grew various short-day plants in prolonged 72-hour dark cycles, and probed the dark period with a flash of light at various points. He found three successive peaks of inhibition of flowering by the light, indicating three successive "sleep" periods, approximately 24 hours apart (Fig. 1-29). Thus it appears that the basic architecture of the plant includes timing devices upon which are superimposed the normal regulatory mechanisms such as photoperiodic control.

A relationship between response to a light flash given at different times during the dark period and an overt diurnal rhythm has recently been shown in *Coleus* (19). The critical day length for flowering in this short day plant is 12–13 hours. This critical photoperiod also affects the timing of the diurnal leaf movements. Thus, the leaves show a minimum leaf position about 5 hours after the onset of the dark period when the photoperiod is under 12 hours, but this minimum is seen after only 3 hours of darkness in photoperiods exceeding 12 hours (Fig. 1-30). The floral induction phase (defined as that period when light inhibits floral develop-

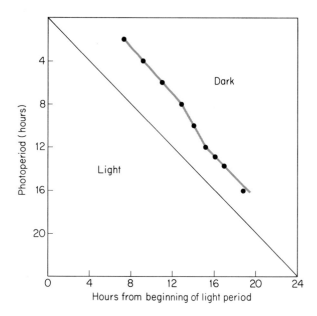

Fig. 1-30   The time of the minimum leaf position of *Coleus frederici* under different day lengths follows the light/dark transition (diagonal line) by 3 hours in day lengths longer than the critical (12–13 hours) but by 5 hours in inductive day lengths shorter than the critical. [From Halaban (18).]

ment) always followed the minimum leaf position by 5 hours; thus, it occurred 10 hours after the onset of darkness in photoperiods below 12 hours and after 8 hours of darkness in photoperiods exceeding 12 hours (Fig. 1-31). This implies that the photoperiod does not control floral induction directly, but rather that it alters the endogenous rhythm, which changes at a photoperiod near the critical for flowering. Thus, the endogenous rhythm in some way governs the flowering response.

The nature of the control appears to reside in an interaction between the states of phytochrome present and the endogenous rhythm. As a flash of red light during the dark period will induce flowering in a long-day plant, it would be thought that a high state of P_{fr} would promote flowering. But when red light is given to *Lolium temulentum* (Darnel grass) to extend a short day, it is ineffective unless far-red light is also present, suggesting that high P_{fr} throughout the dark period is inhibitory (20). The sensitivity of the plant to red and far-red light appears to change throughout a 24-hour cycle. Red light alone is inhibitory to flowering following 8 hours of white light, reaching a maximum inhibition 12 hours after the start of the light phase. Then it becomes promotive again, whereas far-red light operates in reverse, being promotive 12 hours after the beginning of the

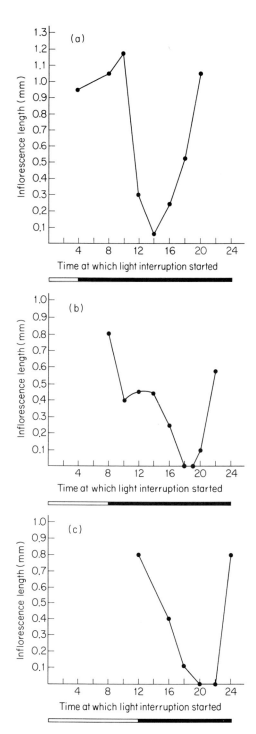

Fig. 1-31 The flowering response of *Coleus frederici* (expressed as the inflorescence length) to 30- or 60-minute light exposures at various times after the onset of darkness under different photoperiods [(a) 4 hours, (b) 8 hours, (c) 12 hours]. The time of maximum inhibition of flowering by light occurs 10 hours after the onset of darkness in inductive photoperiods of less than 12 hours but only 8 hours after darkness at 12 hours, just below the critical day length. It can be seen that under a 14-hour photoperiod, the time at which light interruption of the dark period inhibits flowering would coincide with the next photoperiod. [From Halaban (18).]

light phase. From this it seems that either P_{fr} is first inhibitory, then promotive to flowering, or that P_r is required early in the supplemental light phase. The fact that the sensitivities vary 12 hours after the beginning of the light period clearly indicates an interaction with diurnal rhythms. In *Chenopodium rubrum*, a short-day plant, the addition of far-red light during the light phase prevents flowering in the same way as interrupting the dark phase with red light. This indicates a positive role of P_{fr} during the light phase, again clearly related to rhythmic changes within the plant.

Even the rapid responses of leaf movements in *Albizzia* appear to be influenced by diurnal rhythms. Far-red light has been found to have a greater inhibitory effect on closing in the dark when given early in the light period than when given late in the light period. From all this evidence, it appears that action of phytochrome is closely related to endogenous rhythms. This considerably complicates any explanation of the action of phytochrome on flowering, as the nature of the endogenous rhythm itself remains very mysterious.

Vernalization and low temperature requirement for flowering

An exceptionally interesting interplay of photoperiod and temperature can be seen in biennial plants. These plants are unable to complete their life cycle in one year in the temperate zones and must overwinter before they are ready to receive a photoperiodic stimulus. The crucial factor in the overwintering process is continued exposure to low temperatures of the order of $5\,°C$ for extended periods of approximately 4 to 6 weeks (Fig. 1-32). Only after such an exposure to low temperatures will the plants flower in long days.

The physiological basis for the action of low temperature is still very mysterious. Once again it appears that the hormone gibberellin may substitute, at least in part, for the cold treatment. Gibberellin-treated plants frequently become sensitive to long photoperiods, whereas they would not be sensitive without this treatment (Fig. 1-33). Perhaps the cold period may produce either gibberellin or gibberellin precursors that are then converted to gibberellin in long days. Gibberellins have, in fact, been found to result from the cold treatment of seeds that require a cold period to stimulate germination (see Chapter 6). Alternatively, as we shall see in later discussions, it is generally assumed that gibberellin acts by stimulating the formation of new ribonucleic acid and protein molecules. Thus, low temperature may provide other biochemical substrata upon which phytochrome regulation is dependent and superimposed.

Biennial plants frequently differ from annual plants of the same type

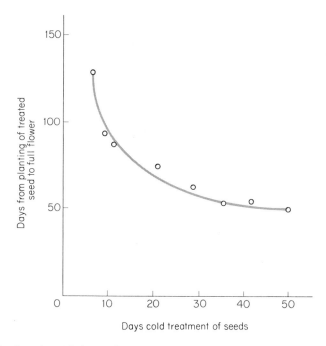

Fig. 1-32   Some biennial plants will not flower at all without a cold treatment. Rye gives a quantitative response, the length of time between planting and flowering decreasing as the period of cold treatment is lengthened. (From Purvis and Gregory. 1937. Ann. Bot. N.S. 1:569–592.)

by a single gene, the annual type generally being dominant. It may, therefore, be supposed that the biennial habit arose by a single gene mutation from an annual ancestor. The difference between the two habits is extremely important in nature and in agriculture in determining the geographical range over which the varieties can grow. This is seen most graphically in the cereals, in which the winter varieties are generally sown late in the summer or early autumn. They germinate immediately to reach the size of a stubble, which is then usually covered by snow for extended parts of the winter, providing the required cold period. The plants then commence growth early in the spring, flower early in response to long days, and are harvested in summer. By contrast, the spring varieties are sown very early in the spring and, not requiring a cold period, flower that summer and are harvested in the autumn of the same year. It can be seen that, depending on the climatic conditions in a given region, it might be desirable to use either the spring or the winter variety. In some parts of the world, the growing season is so short that spring types cannot be used because they mature too late in the fall and are nipped by early fall frosts. In some of these same regions, winter types cannot be used because the winters are so severe that the seedlings could not withstand the low

Fig. 1-33   Carrot plants showing the application of gibberellin as a substitute for a cold requirement for flowering. Treatments: Left, no gibberellin, no cold treatment. Center, 10 μg gibberellin per day, no cold treatment. Right, no gibberellin but 8 weeks cold treatment. The plants were grown in long days. Note that the gibberellin- and cold-treated plants are almost identical. (From Lang. 1957. Proc. Nat. Acad. Sci. **43**:709–717.)

temperatures that exist in the field. Under these conditions, seeds can be allowed to begin germination indoors in early spring and then can be exposed to low temperatures (2 to 5 °C) for a period of 4 to 6 weeks; when the conditions in spring permit sowing, they are planted in the usual fashion. This artificial administration of low-temperature treatments in an attempt to hasten flowering is referred to as *vernalization*. Having been vernalized, the plants are off to a quick start, able to initiate flowers in response to the long days of the late spring and early summer, and are ready for harvest by the late summer, before the first frosts come. Vernalization, which is practiced in agriculture in some parts of the world, furnished the basis for the claims of the Russian agronomist Lysenko to have transformed the genetic nature of plants by environmental treatment. We now know these claims to be false; the progeny of vernalized plants are not, as Lysenko claimed, temperature-independent. They still need low temperatures to initiate flowering, as did the vernalized seed from which they arose.

Photoperiodism in animals

Following shortly on the discovery of photoperiodism in plants, it was demonstrated that many animals, including birds, insects and mammals, are also sensitive to this factor of the environment. The onset of migratory patterns and reproduction in birds, the making and breaking of diapause in insects, and the incitement to oestrus in mammals can, under appropriate conditions, be shown to result from response to day length. Yet there is no evidence in any of these forms of the existence of or control by phytochrome. Where action spectra have been made in animals, they differ rather sharply from those obtained with plants. We can only presume that response to photoperiod arose independently several times in the course of evolution, and that the pigment used by animals for this control is not phytochrome. In common with plants, the photoperiodic stimulus is translated into hormonal production to accomplish the onset of reproduction.

Summary

Many aspects of plant development are controlled by a pigment, phytochrome. Phytochrome is a protein with an open chain tetrapyrrole chromophore and exists in two mutually photoreversible forms, absorbing respectively at maxima of 660 and 730 nm. The latter is presumably the active form, and may exert its control directly by regulation of membrane permeability. Among the physiological processes ultimately under the control of this pigment are seed germination, growth and development of stems and leaves, and the onset of reproduction in many plants. In the control of reproductive patterns, plants respond to the length of the uninterrupted dark period. Because in the 24-hour cycle night length is compulsorily linked to day length, the phenomenon was originally thought to depend on length of day, and hence the name, *photoperiodism*. Plants may either respond to short days (i.e., long nights), or to long days (i.e., short nights), or may not respond at all. The control of flowering by phytochrome is intricately linked with the endogenous rhythms within the plant. Plants may be unresponsive to photoperiod either for genetic reasons, because they are too young (not ripe to flower), or because they first require prolonged exposure to low temperature. Many biennial plants fall into the last pattern. The artificial administration of low temperature in order to hasten reproduction is called *vernalization*. Most plants requiring cold treatment as a prelude to flowering are long-day types. In some plants, administration of gibberellin can satisfy both the requirement for low temperature and long day.

REFERENCES

General

Hendricks, S. B. and H. A. Borthwick. 1967. The function of phytochrome in regulation of plant growth. Proc. Nat. Acad. Sci. U.S. **58**:2125–2130.

Hillman, W. S. 1962. The physiology of flowering. Holt, Rinehart and Winston, New York. 164 p.

Hillman, W. S. 1967. The physiology of phytochrome. Ann. Rev. Plant Physiol. **18**:301–324.

Salisbury, F. B. 1963. The flowering process. Pergamon, Oxford. 234 p.

Salisbury, F. B. 1965. The initiation of flowering. Endeavour **24**:74–80.

Siegelman, H. W. and W. L. Butler. 1965. Properties of phytochrome. Ann. Rev. Plant Physiol. **16**:383–392.

1. Garner, W. W. and H. A. Allard. 1920. Effect of the relative length of day and night and other factors of the environment on growth and reproduction in plants. J. Agr. Res. **18**:553–606.

2. Garner, W. W. and H. A. Allard. 1923. Further studies in photoperiodism, the response of the plant to the relative length of day and night. J. Agr. Res. **23**:871–920.

3. Hamner, K. C. and J. Bonner. 1938. Photoperiodism in relation to hormones as factors in floral initiation and development. Botan. Gaz. **100**:388–431.

4. Hendricks, S. B. and H. A. Borthwick. 1954. Photoperiodism in plants. Proc. 1st Intern. Photobiol. Congr., pp. 23–35.

5. Flint, L. H. and E. D. McAlister. 1937. Wavelengths of radiation in the visible spectrum promoting the germination of light-sensitive lettuce seed. Smithsonian Inst. Misc. Collections **96(2):**1-8.

6. Borthwick, H. A., S. B. Hendricks, M. W. Parker, E. H. Toole and V. K. Toole. 1952. A reversible photoreaction controlling seed germination. Proc. Nat. Acad. Sci. U.S. **38:**662-666.

7. Butler, W. L., S. B. Hendricks and H. W. Siegelman. 1965. Purification and properties of phytochrome, p. 197-210. *In* T. W. Goodwin [ed.] The chemistry and biochemistry of plant pigments. Academic Press, New York.

8. Mumford, F. E. and E. L. Jenner. 1966. Purification and characterization of phytochrome from oat seedlings. Biochemistry **5:**3657-3662.

9. Mohr, H. 1966. Differential gene activation as a mode of action of phytochrome 730. Photochem. Photobiol. **5:**469-483.

10. Fondéville, J. C., H. A. Borthwick and S. B. Hendricks. 1966. Leaflet movement of *Mimosa pudica* L. indicative of phytochrome action. Planta **69:**357-364.

11. Jaffe, M. J. 1968. Phytochrome-mediated bioelectric potentials in mung bean seedlings. Science **162:**1016-1017.

12. Galston, A. W. 1968. Microspectrophotometric evidence for phytochrome in plant nuclei. Proc. Nat. Acad. Sci. U.S. **61:**454-460.

13. Haupt, W. 1965. Perception of environmental stimuli orienting growth and movement in lower plants. Ann. Rev. Plant Physiol. **16:**267-290.

14. Koukkari, W. L. and W. S. Hillman. 1968. Pulvini as the photoreceptor in the phytochrome effect on nyctinasty in *Albizzia julibrissin.* Plant Physiol. **43:**698-704.

15. Hodson, H. K. and K. C. Hamner. 1970. Floral inducing extract from *Xanthium.* Science **167:**384-385.

16. Reid, D. M., J. B. Clements and D. J. Carr. 1968. Red light induction of gibberellin synthesis in leaves. Nature **217:**580-582.

17. Bewley, J. D., M. Black and M. Negbi. 1967. Immediate action of phytochrome in light-stimulated lettuce seeds. Nature **215:**648-649.

18. Cumming, B. G. and E. Wagner. 1968. Rhythmic processes in plants. Ann. Rev. Plant Physiol. **19:**381-416.

19. Halaban, R. 1968. The flowering response of *Coleus* in relation to photoperiod and the circadian rhythm of leaf movement. Plant Physiol. **43:**1894-1898.

20. Vince, D. 1965. The promoting effect of far-red light on flowering in the long day plant *Lolium temulentum.* Physiol. Plant. **18:**474-482.

Two

Ethylene

The old observation that one rotten apple in a barrel causes the entire lot to spoil can now be explained in simple scientific terms. The one rotten apple produces a volatile agent, ethylene (C_2H_4), which causes necrotic changes in the healthy fruits nearby. This leads to their spoilage, and they, in turn, begin to produce ethylene, which affects still other fruit. Thus, there is a cascade effect and a small amount of ethylene can produce a very large result. The well-known and widespread practice of preserving storage apples by enriching the air with carbon dioxide is based on the fact that CO_2 prevents ethylene from exerting its effect.

Although these rather simple facts have been known for a very long time, the action of ethylene on plant cells and the countereffect of carbon dioxide were not regarded as a part of normal regulation in the plant; they were regarded as being related to plant pathology rather than to plant physiology. The ethylene production was thought to result from invasion of the plant by some pathogen, or possibly to the physiological breakdown of some plant cells that results from bruising, unfavorable storage temperatures, or simply aging. In recent years, it has become clear that ethylene is a normal plant metabolite, produced by healthy cells and probably exerting normal regulatory control over such morphogenetic phenomena as de-etiolation, floral initiation and fruit ripening. Since ethylene is produced in minute quantities and may be active in cells other than its site of production, it may legitimately be referred to as a plant hormone.

40

Abnormal growth responses to ethylene applications

Applications of ethylene to etiolated peas have very pronounced effects on growth. Under optimal conditions seedlings of such pea varieties as Alaska grow straight and tall when cultivated in complete darkness. At the age of 7 days, such seedlings consist of an erect, slender, unpigmented stem, a sharply recurved apical hook and a slightly yellow terminal bud. If a stream of air containing ethylene is blown over these seedlings their longitudinal extension is inhibited, and there is a promotion of lateral growth, resulting in a swelling just below the apical bud. In addition, the stems lose their normal sensitivity to gravity, becoming diageotropic or ageotropic (Fig. 2-1). We shall see later that the normal tropistic orientation of stems and roots is at least in part a consequence of the transverse migration of another plant growth hormone, auxin, in response to various unilateral stimuli, such as gravity and light. Ethylene apparently acts to interfere with this normal lateral transport of auxin, thus making impossible the righting reactions that result in the normal geotropic orientations with which we are familiar.

Fig. 2-1   The effect of ethylene applications to etiolated pea seedlings (a) 6 and (b) 18 days old and photographed about 2 weeks after treatment. Note the swelling, lack of extension growth in the growing tissue and loss of sensitivity to gravity. Compare with Fig. 2-4. (From Borgström. 1939. The transverse reactions of plants. Lund, Sweden.)

(a)   (b)

The effects of ethylene on light-grown plants are no less striking. In many plants, e.g., tomato, a pronounced epinasty (downward bending) of the leaves results, owing to a lateral swelling of the cells on the upper side of the base of the petiole. One may also note the development of large numbers of adventitious roots on the stem. These adventitious roots result from a stimulation of cell division in the region of the interfascicular cambium to form a root meristem, which then grows out of the stem as a root. Thus ethylene not only interferes with cell extension, causing lateral swelling, but also influences cell division.

Detection and assay of ethylene

For many years, ethylene was detected and measured quantitatively by a bioassay technique, i.e., by the response of an organism sensitive to ethylene. Most frequently, the assay used was the "triple response" of etiolated peas, namely, cessation of extension growth, swelling and loss of sensitivity to gravity. These responses are so striking that they may be related quantitatively to the concentration of ethylene in the gas stream. Thus, by successive dilution of a sample of gas containing an unknown quantity of ethylene and by reference to the response of etiolated peas, one may obtain some impression of the ethylene content of the gas.

There are several things wrong with this assay. In the first place, it is rather slow. It requires growing pea plants in a dark room for 7 days under conditions rather rigorously excluding ethylene and then waiting another day for results. Second, it is not entirely specific. Other unsaturated hydrocarbons, such as propylene, butylene and acetylene can mimic the effect of ethylene on etiolated peas. Unless these gases are separated before injection into the chamber containing the pea plants, it is impossible to disentangle their effects. Clearly, a more rapid and more specific assay would be preferred.

Early chemical assays for ethylene involved measurement of the amount of bromine added to the double bond of ethylene when the gas was passed through bromine water (Fig. 2-2). This procedure can be made titrimetric and therefore rapid, but it also suffers from lack of specificity, since bromine will add at the double bond of virtually all unsaturated hydrocarbons. A procedure that combines easy separation with quantitative estimation is gas chromatography. In this procedure a mixture of gases

Fig. 2-2 Ethylene can be estimated by combination with bromine to form ethylene dibromide.

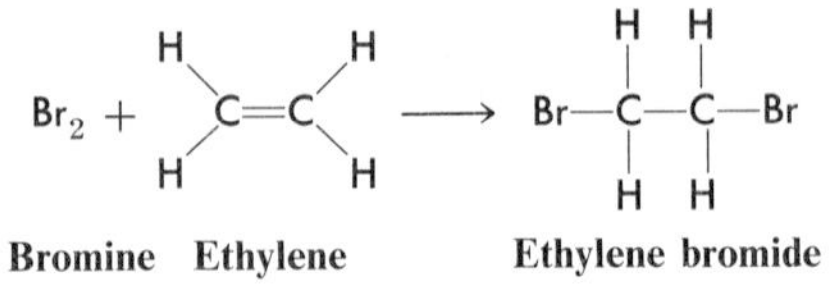

Bromine Ethylene   **Ethylene bromide**

is introduced into a long glass column packed with an inert filler to which is adsorbed some material that differentially restrains the migration of the various gases, producing very sharp separations. As the gas moving through the column comes to the exit port, some kind of detection device is activated, and the magnitude of the activation reflects the amount of gas delivered. Most modern detectors are flame ionization detectors; they are sensitive down to one part per billion of ethylene, a concentration well below that which causes physiological effects. They are easily connected so as to yield automatic records of the number of eluted peaks and

Fig. 2-3 Recorder tracing from a gas chromatographic separation and estimation of ethylene on an activated alumina column at 45°C, with a flame ionization detector. A, air; B, ethane; C, ethylene; D, propane. In this sample 1.2 nanoliters of ethylene were detected in a 0.5 ml injection volume = 2.4 ppm. (Courtesy of G. D. Blanpied, Cornell University.)

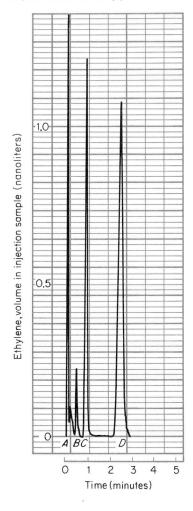

Time (minutes)

the magnitude of each peak (Fig. 2-3). Through the use of such devices, much quantitative information has recently been obtained about the rate of production and ethylene content of many normal plant tissues.

The role of endogenous ethylene in plants

In addition to distorting the growth of plants, application of ethylene may simulate normal morphogenetic events. It is now evident that many of these effects are in fact the result of natural ethylene production by plants.

Etiolation

Plants growing in the darkness differ greatly from normal plants grown exposed to light. Their stems are highly elongated; their leaves are poorly developed; the apex of the stem just below the terminal bud is recurved into an apical hook; they are devoid of pigmentation, and they lack certain internal structural differentiation characteristic of light-grown stems (Fig. 2-4). At least some of these characteristics are results of high ethylene production. For example, in the etiolated pea, it is known that the apical hook forms in response to high ethylene concentrations (1) found primarily in the hook and first node. This high ethylene production is inversely related to the capacity of these tissues to destroy auxin, and indicates that ethylene production may be a result of high auxin concentrations (see next chapter). The opening of the hook in response to red irradiation may be directly correlated with decreased ethylene production, which occurs very shortly after absorption of the red light. Plants so exposed can be caused to reform a hook by the application of large quantities of exogenous ethylene. The exogenous ethylene does not prevent the other symptoms of de-etiolation, such as inhibition of stem growth and promotion of leaf growth, so that one must assume it is not the total answer to the de-etiolation process.

Floral initiation

It has been known for many years that certain species of plants, such as the pineapple, can be induced to flower by the application of unsaturated hydrocarbons, including ethylene (2). This discovery was fortuitously made as a result of the action of illuminating gas, which contains ethylene, upon pineapple plants. Premature flowering induced by the gas could be mimicked by various unsaturated hydrocarbons, including acetylene and ethylene.

Fig. 2-4 An etiolated, dark-grown pea (left) is very different from its light-grown counterpart (right). It is tall and slender with a recurved terminal bud and only scale leaves. The change from an etiolated to a light-grown condition is brought about partly by red light acting on phytochrome. This increases leaf and decreases stem growth, and causes the unfolding of the apical hook. The latter is related to a red-light-induced decrease in ethylene production by the hook.

At first this was regarded as an aspect of abnormal behavior induced by an unnatural external agent. With the discovery that ethylene is a normal plant metabolite, attention has naturally been paid to the possibility that the control of flowering by phytochrome may be somehow linked to ethylene metabolism, just as the control of unhooking of etiolated seedlings by the P_{fr} form of phytochrome is apparently correlated with ethylene metabolism. Although this subject cannot be treated definitively at this time because of a lack of data, there have been experiments in which the interruption of a long dark period in short-day plants by a flash of red light has led to a decrease in ethylene production. Auxin application, which also causes flowering in pineapple, has been shown to exert its effect via ethylene production. It should be noted, however, that ethylene is probably not identical with the flowering hormone, and floral induction by ethylene is by no means a universal phenomenon.

Abscission

Just as illuminating gas can alter the flowering behavior of photoperiodically sensitive plants, so high concentrations of uncombusted illuminating gas can lead to widespread abscission of leaves. This effect can also be traced to the vapors of unsaturated hydrocarbons, including ethylene, found in the gas. Recent evidence suggests that ethylene may indeed be part of the natural abscission process also. We shall discuss the hormonal controls of abscission later in the chapter on senescence.

Fruit ripening

We have already mentioned that the elaboration of ethylene by spoiled, infected or overripe fruits causes the ripening of other nearby fruits so that they in turn produce ethylene, which affects still other fruits. It will be worthwhile to examine both of these processes further.

The ripening of a fruit is a very complex process. The fruit is the enlarged ovary of a flower. Subsequent to pollination and the growth of the pollen tube down into the ovary, the process of double fertilization in the embryo sac of the ovule begins the development of both the embryo and the endosperm. The developing ovule, which is destined to become a seed, then in turn becomes a source of stimulatory material causing the onset of proliferative activity in the wall of the ovary. We shall see in later chapters that at least a part of this proliferation can be mimicked in some fruits by the exogenous application of the two plant growth hormones, auxin and gibberellin. The fruit continues to grow mainly by an increase in cell size. In most cases the cell division phase of fruit growth ends while the fruit is but a small fraction of its ultimate size. It is interesting to note that in figs, not only does ethylene induce ripening, but exogenous applications during the cell growth stage also lead to a considerable increase in cell and fruit size (3).

Once the fruit has attained its maximum size, other more subtle chemical changes begin which ultimately cause it to "ripen" and become edible.

We know, for example, that many a full-grown apple is inedible because of acidity and hardness. The ripening process in the apple consists in part of the disappearance of a large part of the malic acid, which makes the unripe fruit acid. This is followed by a weakening of the cell walls, so that the mechanical stress required to separate one cell from another is greatly diminished. Many fruits ripen more rapidly when they are picked, indicating that the signal for ripening arises in the fruit itself, or that a suppression of ripening is exerted by the other plant parts.

Ripening is closely connected with respiration of the fruit. If one measures the carbon dioxide output from a fruit or a fruit slice during the course of ripening, one notices frequently a rather sharp inflection point denoting greatly increased output of CO_2 for a short period, followed again by a sharp decline. This period of greatly increased carbon dioxide output is called the *climacteric* (Fig. 2-5). Following onset of the climacteric, the fruit rapidly goes through those changes that transform it from a fully

Fig. 2-5   Measurements of carbon dioxide evolution in ripening fruit show a sharp rise in the respiration rate as the fruit ripens. This is termed the *climacteric*. The respiration level then falls as the fruit tissues senesce. (From Biale. 1950. Ann. Rev. Plant Physiol. 1:183–206.)

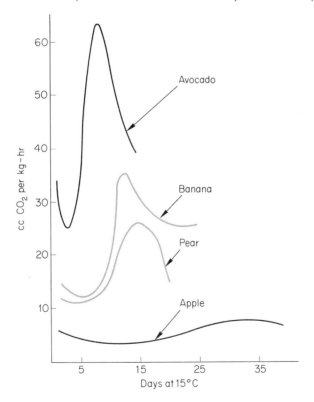

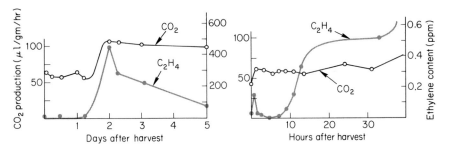

Fig. 2-6   The relationship between respiration and the endogenous ethylene content in avocado. Left, gross pattern. Right, the ethylene scale is magnified to reveal small changes that occur at the onset of ripening. (From Burg and Burg. 1962. Nature **194**:398–399.)

grown unripe fruit to a ripe fruit ready to be eaten. The climacteric, once thought to be a degenerative process, can be shown to be a positive change requiring the expenditure of energy. It is thus prevented by inhibitors of respiration, by high carbon dioxide or nitrogen concentrations, or by low temperature. Application of ethylene, by contrast, promotes both the climacteric rise and ripening in mature fruit. Originally, investigations on the role of endogenous ethylene production during ripening indicated that ethylene production commenced after the climacteric. With the advent of more refined techniques such as gas chromatography, however, more precise measurements showed that ethylene is consistently produced at or before the onset of the climacteric (Fig. 2-6). A little ethylene is, in fact, present all the time, but this is boosted about a hundredfold at the climacteric. Even in fruit where very little ethylene is produced, e.g., the mango, this can still be sufficient to cause the rise in respiration (Fig. 2-7). When ripening is prevented by such conditions as low temperature, ethylene production is also inhibited. It may, therefore, be concluded that ethylene is the natural fruit-ripening hormone in plants. Further evidence to support this conclusion has been obtained by removing the ethylene from the fruits as fast as it is formed by placing the fruit at a reduced pressure but maintaining the oxygen concentration similar to the atmosphere. Under these conditions ripening is delayed.

How does the ethylene induce ripening? Although in some fruits (e.g., avocado and mango) ethylene production and the increase in respiration run parallel, in others, such as the banana, ethylene only acts as a trigger, the production of ethylene declining before the maximum rate of respiration is achieved (Fig. 2-8). This indicates that ethylene must cause the onset of some other process, which then brings about ripening.

There are two current theories to explain the biochemical changes that occur in ripening (4) and are presumably, therefore, triggered by ethylene.

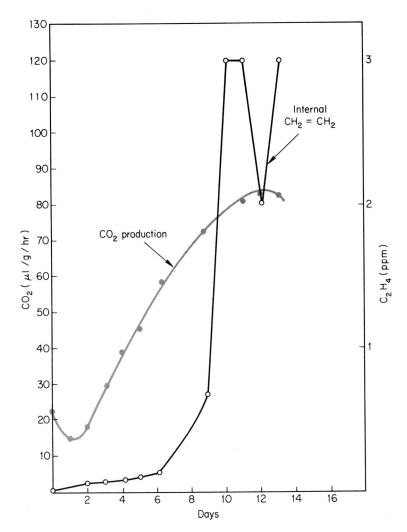

Fig. 2-7 The endogenous ethylene content in ripening mangoes is very low compared to other fruits (cf. avocado, Fig. 2-6), but a small increase in ethylene content is still noticeable prior to the rise in CO_2 output. (From Burg and Burg. 1962. Plant Physiol. **37**:179–189.)

It is over 40 years since respiration changes during ripening were first explained in terms of a change in "organization resistance" or the separation of enzymes and substrate by semipermeable membranes (5). This hypothesis has been revived in recent years by studies on tissue permeability. Permeability changes occur in the cellular membranes preceding or during ripening, and this is indicated by a leakage of solutes from the

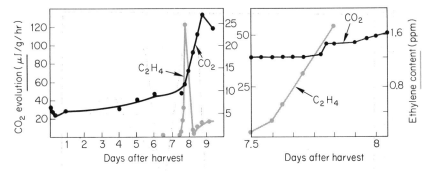

Fig. 2-8   The relationship between respiration and the endogenous ethylene content in banana. Left, gross pattern. Right, the ethylene scale is magnified to reveal small changes that occur at the onset of harvest. (From Burg and Burg. 1965. Bot. Gaz. **126**:200–204.)

cells (6). In banana, for example, an increase in permeability was detected 44 hours before the climacteric and rose rapidly thereafter, so that during the climacteric the tissue was maximally permeable. Additional evidence

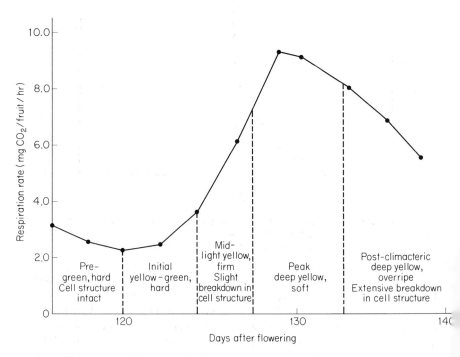

Fig. 2-9   Correlation of changes in color, texture and cell structure with changes in respiration during the climacteric in ripening Williams pears. [From Bain and Mercer (7).]

has come from electron-microscope investigations in pears during ripening (7). Here the respiratory changes appear to be correlated with a structural breakdown of the cell membranes (Fig. 2-9), though not of the mitochondria, which retain their capacity for phosphorylation and synthesis during the climacteric. With the breakdown of membrane permeability the release of enzymes could cause a speedup of the metabolic processes, particularly those associated with ripening, namely, respiration and breakdown of acids and wall constitutents. Consistent with its role in initiating ripening, ethylene has also been found to increase tissue permeability (8).

The other theory derives from the discovery that there is an increase in protein content during the climacteric. This emphasizes the view that ripening is an active process; it is suggested that the increase in protein synthesis during ripening is related to the formation of the enzymes required for the various biochemical changes occurring during the ripening process (4, 9). If mature fruit tissues are treated with inhibitors of protein synthesis, ripening is prevented parallel with the reduction in protein synthesis (Fig. 2-10). Ethylene synthesis itself is dependent upon protein synthesis at the early climacteric stage, but less so once the climacteric has progressed (Fig. 2-11). Ethylene has been shown to induce protein formation in several tissues, while in the absence of protein synthesis the addition of external ethylene does not induce ripening. Thus, ethylene appears to have a dual involvement with protein synthesis during ripening, first as an initiator or effector of specific protein synthesis, and second as a product of the ripening metabolism involving the same proteins. This provides a basis for the chain reaction in the barrel of apples!

Does ethylene act, therefore, on permeability or protein synthesis? This is not clear at present. Perhaps it causes both independently, and both contribute to ripening. Alternatively, first the permeability could be selectively increased by ethylene, and this might then lead to an increased synthesis of the enzymes involved in ripening.

The biochemical origin of ethylene

There is considerable evidence that ethylene is derived from carbon atoms 3 and 4 of the amino acid, methionine (10) (Fig. 2-12). This has been demonstrated by feeding labeled methionine to plant tissues and noting the fate of the various carbon atoms. Methionine, and especially its derivative, methional, can also give rise to ethylene in vitro under the action of either the enzyme peroxidase, or riboflavin mononucleotide in the presence of light. Both of these model systems may have physiological importance, for, if peroxidase makes ethylene, and peroxidase is among the enzymes induced by ethylene (11), then the induction of peroxidase

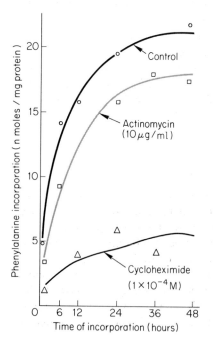

Fig. 2-10(a)   The influence of cycloheximide (10 μg/ml) treatment on flesh softening of Bartlett pears at the midclimacteric stage. The cycloheximide, which is an inhibitor of protein synthesis, was infiltrated in the presence of mannitol, which was also used alone.

Fig. 2-10(b)   The influence of actinomycin-D, an inhibitor of RNA synthesis, and cycloheximide on incorporation of ^{14}C-phenylalanine into protein of midclimacteric Bartlett pears. [From Frenkel, et al. (9).]

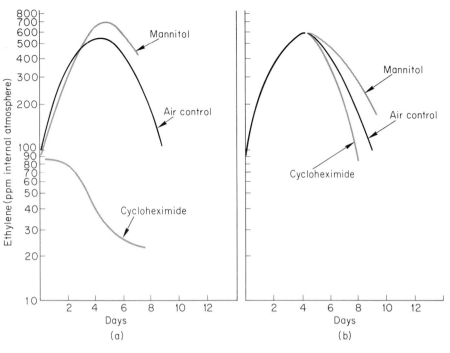

Fig. 2-11   The influence of cycloheximide (10 μg/ml) treatment on ethylene synthesis of Bartlett pear fruits at (a) early and (b) peak climacteric stages. The cycloheximide was infiltrated in mannitol, which was also used alone. [From Frenkel, et al. (9).]

Fig. 2-12   Schematic pathway for the formation of ethylene from methionine. Peroxidase and various cofactors are required, depending upon the system under consideration. The numerals indicate the origin and fate of the carbon atoms in methionine.

Possible cofactors and requirements

Flavine mononucleotide and light

Peroxidase and H_2O_2

Metal ions

$NH_3 + \overset{1}{C}O_2$

$\overset{5}{C}H_3SH + \overset{2}{H}COOH$

$H_3-S-\overset{4}{C}H_2-\overset{3}{C}H_2-\overset{2}{C}H-\overset{1}{C}OOH \longrightarrow \overset{5}{C}H_3-S-\overset{4}{C}H_2-\overset{3}{C}H_2-\overset{2}{C}HO \longrightarrow \overset{4}{C}H_2=\overset{3}{C}H_2$

NH_2

Methionine   **Methional**   **Ethylene**

by ethylene could be the means for propagating the production of ethylene from one tissue to another. Not only do peroxidases produce ethylene, but some isozymes of peroxidase are more active in this regard than others. This may be the method by which auxin promotes ethylene formation (see Chapter 4), as auxin has been found to regulate the synthesis of certain isoperoxidases. Similarly, the production of ethylene under the action of light and riboflavin may have some implications for phototropism. Since ethylene is known to block longitudinal transport of auxin, the local accumulation of ethylene on the lighted side of a unilaterally illuminated cylindrical organ could, in fact, be the agency whereby the transport of auxin is hindered on the lighted side, leading to the diminution of growth there and the subsequent curvature. We shall explore this matter again in the next chapter on tropisms.

Summary

The simple hydrocarbon ethylene, C_2H_4, is a normal plant metabolite that can control the process of hook formation in etiolation, the onset of flowering in some plants, the abscission of leaves, the induction of the climacteric period of fruit respiration, and the subsequent processes related to fruit ripening. Ethylene appears to induce the onset of the climacteric, which is an active, energy-requiring process, by causing an increase in the synthesis of the enzymes involved in ripening. This may be preceded by ethylene-induced increases in membrane permeability. Exogenous applications of ethylene may also cause lateral swelling of cells, the general loss of polarity associated with ageotropic behavior and the stimulation of cell division that results in the formation of adventitious roots. Ethylene is presumably derived metabolically from the amino acid, methionine. Model systems that produce ethylene from methionine include peroxidase and also riboflavin in the presence of light. Both of these systems may be related to the normal physiological processes, such as the propagation of the ethylene effect in ripening fruits and the curvature of unilaterally illuminated plant organs toward light.

REFERENCES

General

Burg, S. P. and E. A. Burg. 1965. Ethylene action and the ripening of fruits. Science **148:**1190–1196.

Dilley, D. R. 1969. Hormonal control of fruit ripening. Hortscience **4:**111–114.

Hansen, E. 1966. Postharvest physiology of fruits. Ann. Rev. Plant Physiol. **17**:459-477.

Mapson, L. W. 1970. Biosynthesis of ethylene and the ripening of fruit. Endeavour **29**:29-33.

Pratt, H. K. and J. D. Goeschl. 1969. Physiological roles of ethylene in plants. Ann. Rev. Plant Physiol. **20**:541-584.

1. Kang, B. G., C. S. Yocum, S. P. Burg and P. M. Ray. 1967. Ethylene and carbon dioxide: Mediation of hypocotyl hook-opening response. Science **156**:958-959.

2. Burg, S. P. and E. A. Burg. 1966. Auxin induced ethylene formation: its relation to flowering in pineapple. Science **152**:1269 and references therein.

3. Maxie, E. C. and J. C. Crane. 1968. Effect of ethylene on growth and maturation of the fig, *Ficus carica* L., fruit. Proc. Am. Soc. Hort. Sci. **92**:255-267.

4. Hulme, A. C., M. J. C. Rhodes, T. Galliard and L. S. C. Wooltorton. 1968. Metabolic changes in excised fruit tissue. IV. Changes occurring in discs of apple peel during the development of the respiration climacteric. Plant Physiol. **43**:1154-1161.

5. Blackman, F. F. and P. Parija. 1928. Analytic studies in plant respiration I, II, III. Proc. Roy. Soc. B. **103**:412-523.

6. Sacher, J. A. 1966. Permeability characteristics and amino acid incorporation during senescence (ripening) of banana tissue. Plant Physiol. **41**:701-708.

7. Bain, J. M. and F. V. Mercer. 1964. Organization resistance and the respiration climacteric. Aust. J. Biol. Sci. **17**:78-85.

8. Von Abrams, G. J. and H. K. Pratt. 1967. Effect of ethylene on the permeability of excised cantaloupe tissue. Plant Physiol. **42**:299-301.

9. Frenkel, C., I. Klein and D. R. Dilley. 1968. Protein synthesis in relation to ripening of pome fruits. Plant Physiol. **43**:1146-1153.

10. Yang, S. F. 1969. Biosynthesis of ethylene, p. 1217-1228. *In* F. Wightman and G. Setterfield [eds.] Biochemistry and physiology of plant growth hormones. Runge Press, Ottawa.

11. Hall, W. C. and P. W. Morgan. 1964. Auxin-ethylene interrelationships, p. 728-745. *In* Régulateurs naturels de la croissance végétale. C.N.R.S., Paris.

THREE

Auxin and Tropisms

Place a plant on its side, and its growth pattern is soon altered. The roots curve down, pointing toward the center of the earth; the stem curves upward, pointing away from the center of the earth; the petioles of the leaves readjust their positions so that the laminae are oriented parallel to the surface of the earth. Each of these changes has a built-in survival advantage for the plant. The roots, which anchor and support the plant and absorb water and nutrients from the soil, obviously function best by growing directly downward. The stem, which supports the leafy photo-synthetic organs, serves the plant best if it elevates the leaves away from the surface of the earth, where there are less likely to be impediments to the direct absorption of light. Finally, the leaves, the receptors of the light quanta in photosynthesis, obviously function most efficiently in light absorption if their broad surfaces are oriented perpendicular to the sun's rays. These growth responses to the unidirectional gravitation stimulus are referred to as *geotropism*. The root is said to be positively geotropic, the stem negatively geotropic, and the leaves plagiogeotropic or diageotropic (at an angle or right angle to gravity, respectively).

The plant's responses to gravity are modified by responses to other environmental stimuli. For example, it is well-known that the direction of growth of stems, roots and leaves is influenced by light as well as gravity. In this phenomenon of plant growth alteration by unilateral light, referred to as *phototropism*, the stems are generally positively phototropic, roots generally negatively phototropic and leaves once again plagiotropic. In both photo- and geotropism the curvature is the result of differential growth on the two sides of the plant axis. The side toward which the curvature occurs grows less rapidly than the side opposite and the resultant of this differential growth is the curvature. Other stimuli that produce tropistic responses are mechanical (thigmotropism), electricity (electrotropism), magnetism (magnetotropism) and chemical substances (chemotropism).

For each tropism there are at least three questions that need to be analyzed.

1. What is the mechanism by which the stimulus is perceived?
2. What is the mechanism by which the very small stimuli become amplified so as to direct the growth of the plant for a long time, even after the stimulus has been removed?
3. How does the amplified stimulus directly control growth?

Although the answers to the first two questions differ for each of the tropisms studied, the answer to the third question seems to be general to several, if not all, of the tropisms. It will, therefore, be useful for us to examine the third question first.

Phototropism and the discovery of auxin

If a dark-grown seedling of a grass is exposed to unilateral light, the coleoptile or leaf sheath rapidly bends toward the light. The major curvature occurs a short distance behind the apex. If one probes with a fine beam of light or covers part of the coleoptile with an opaque barrier to determine what regions of the coleoptile are most sensitive to the light energy, then it appears that light absorbed by the extreme apex is maximally effective (Fig. 3-1). Charles Darwin, who first noticed this, hypothesized that there must be some connecting link between the extremely photosensitive tip and the cells at some distance from the apex, which responded to the stimulus. This hypothesis was sustained when it was shown, many years later, that the apex of the coleoptile, when excised and placed on a moist medium, such as a block of gelatin or agar, released into the medium substantial quantities of a growth-promoting material, named *auxin.*

The growth of the coleoptile cells below the tip is controlled by auxin produced in the tip. Thus, surgical removal of the tip leads to a diminished growth rate of the cells in the decapitated stump, and replacement of the tip leads to a resumption of the original growth rate. If the excised tip

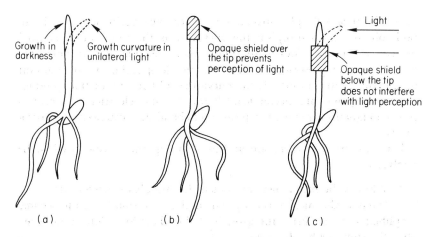

Fig. 3-1 Coleoptiles will grow towards unilateral light (a). If the tip of the coleoptile is covered with aluminum foil, there is no response to light (b), while darkening the area in which the growth occurs is without effect (c). The tip must thus receive the stimulus and transmit it to the growing zone of the coleoptile below the tip.

is permitted to stand on a bloc of gelatin or agar for several hours and then discarded, the block acts as if it contains the auxin originally present in the tip, for if placed symmetrically on the cut stump, it restores the growth rate (Fig. 3-2). When the block is placed asymmetrically on the

Fig. 3-2 Experiments to demonstrate the production of growth-promoting substances by the stem apex.

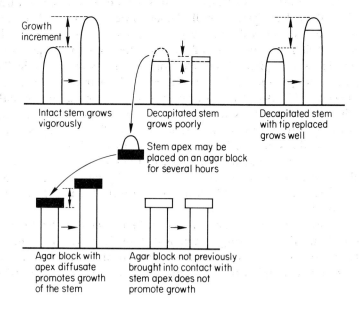

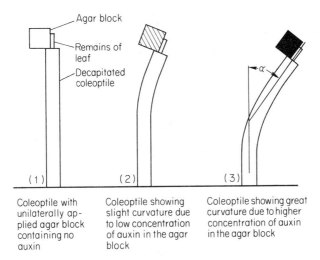

Agar block

Remains of leaf

Decapitated coleoptile

α

(1) | (2) | (3)

Coleoptile with unilaterally applied agar block containing no auxin

Coleoptile showing slight curvature due to low concentration of auxin in the agar block

Coleoptile showing great curvature due to higher concentration of auxin in the agar block

Fig. 3-3   When an agar block containing auxin is placed asymmetrically on a decapitated oat coleoptile, it produces growth of the coleoptile only on the side on which it was placed, resulting in curvature. The bending of the coleoptile can be used as a bioassay to determine the auxin concentration in the agar block. The angle α is related to the concentration in the block.

coleoptile stump, it promotes growth only on that side with which it is in contact (Fig. 3-3), resulting in a curvature away from the side of its application. This discovery led to the hypothesis that light could produce curvature by causing an asymmetry in the distribution of growth substance along the two sides of the coleoptile. This theory was, in turn, confirmed when it was shown that following illumination, a coleoptile tip placed astride two agar blocks separated by a barrier diffuses significantly greater quantities of auxin into the block under the shaded side than into that under the illuminated side (Fig. 3-4). In many experiments the ratio of auxin on the dark side to that on the light side was about two to one.

For many years the chemical identity of auxin remained in doubt, but since about 1940 it has been clear that the major auxin in most plant tissues is the very simple chemical substance indole-3-acetic acid (IAA) (Fig. 3-5). This substance can be collected from tissues either by diffusion or by extraction with organic solvents such as diethyl ether or chloroform. When applied back to plant tissue, the extracted IAA produces a variety of responses, including enhancement of elongation of coleoptiles and stems and, if applied unilaterally, induction of curvature in various organs. Both the promotion of growth and the induction of curvature can be used as bioassays for auxin (see Fig. 3-3). The bioassay determines the amount of active auxin present in an extract by noting the response of an organism to the extract, compared to the response produced by known amounts of auxin. Thus, the curvature of an oat coleoptile unilaterally supplied with

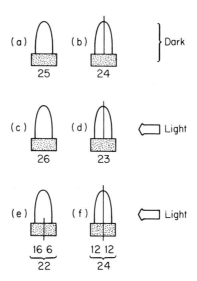

Fig. 3-4   Unilateral light causes the redistribution of auxin as it is transported down from the apex of a coleoptile. The auxin was collected in agar blocks and the amounts from split or intact coleoptiles were measured by the coleoptile curvature test. (The figures indicate the relative amounts of auxin recovered.) Light has not affected the total amount of auxin recovered, but when lateral transport of auxin is permitted in unilateral light by splitting the lower part of the coleoptile and agar block only (e), more auxin is found on the side away from the light. (From Briggs, et al. 1957. Science **126:**210–212.)

auxin in an agar block may be used as a means of determining the quantity of auxin in the block, provided that a "standard curve" for curvature response to pure IAA is also available.

The quantity of indoleacetic acid in tissue may also be assessed by direct physical or chemical measurements. For example, IAA shows an absorption maximum near 280 nm in the ultraviolet region of the spectrum. Thus a suitably purified extract from a plant can be assayed for IAA merely by determining the height of this peak. IAA also develops colors with various reagents, such as acidified ferric chloride (Salkowski's reagent), or *p*-dimethylaminobenzaldehyde (Ehrlich's reagent). The intensity of the

Fig. 3-5   The structural formula indole-3-acetic acid (IAA).

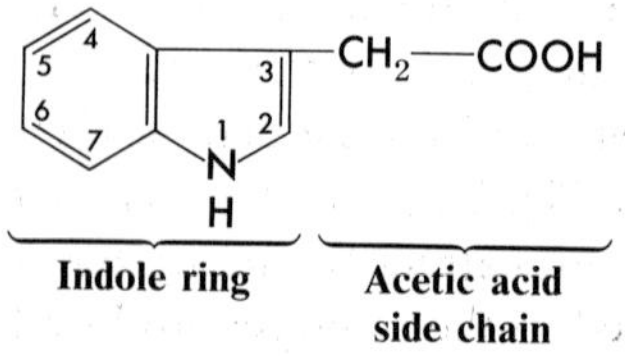

$$\text{—CH}_2\text{—COOH}$$

Indole ring   **Acetic acid side chain**

color developed either in solution or on a chromatogram can be used as a measure of IAA. By the use of any of these techniques, it can be demonstrated that tropistic curvatures are invariably correlated with an unequal distribution of auxin on the two sides of the unilaterally stimulated organ. Thus at least part of the answer to the physiological puzzle posed by tropistic curvatures would be found in an explanation for the induced asymmetry in auxin distribution. Among the explanations that have been advanced are asymmetry in auxin production at the very apex, lateral transport of symmetrically produced auxin, asymmetric destruction of auxin and asymmetrically inhibited transport of auxin. Most researchers now agree that lateral transport of indoleacetic acid occurs under tropistic stimulation, and indeed data from the use of ^{14}C-labeled applied IAA bear out this conclusion (1). The mechanism through which a unilateral stimulus of light or gravity becomes transduced into lateral auxin transport is still unclear, but it possibly involves the development of transverse bioelectric potentials under the influence of aerobic metabolism.

The actions of auxin in the regulation of plant growth

Once auxin had been identified as indole-3-acetic acid, an easily synthesized compound, it was inevitable that many physiologists would apply it to plants in various ways in an attempt to understand how it acts. Such experiments have led to an amazingly large catalog of effects induced by auxin and by synthetic compounds analogous to auxin in structure and function. Whether in fact auxin normally controls those processes that artificially large external applications of this compound can be shown to influence is not yet clear.

Indoleacetic acid appears to be derived in the plant from the amino acid tryptophan through various enzymatic steps involving oxidative deamination to form indolepyruvate, decarboxylation to form indoleacetaldehyde and finally oxidation of the terminal aldehyde group to form IAA (Fig. 3-6). Alternatively, decarboxylation to tryptamine may occur first, followed by oxidative deamination to form indoleacetaldehyde, and then a similar conversion to IAA. There also exist enzymes in the plant, including peroxidase and phenol oxidases, which under proper conditions can destroy IAA, yielding such inactive products as methyleneoxindole and indole-3-aldehyde. These enzymes probably work only under conditions where the auxin concentration rises several orders of magnitude above its usual physiological level. There are also detoxification reactions in which IAA is conjugated with various materials, such as aspartic acid, inositol, and D-glucose and sequestration reactions, involving the union of IAA with various macromolecules.

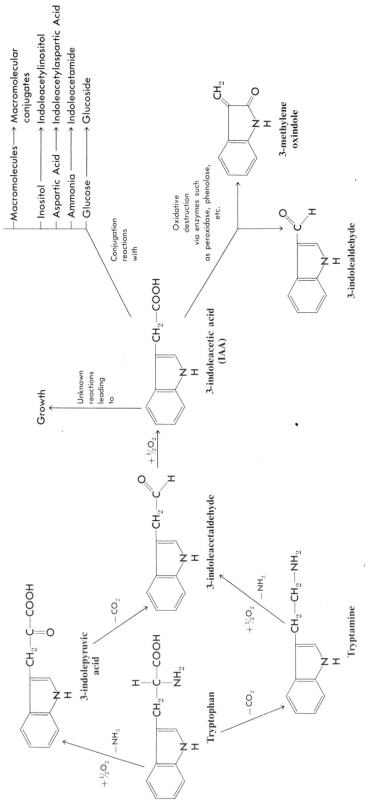

Fig. 3-6 The formation and metabolism of indole acetic acid.

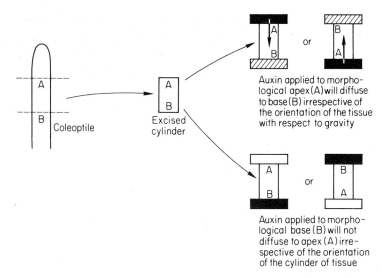

Auxin applied to morpho-
logical apex(A)will diffuse
to base(B)irrespective of
the orientation of the tissue
with respect to gravity

Auxin applied to morpho-
logical base(B)will not
diffuse to apex(A)irre-
spective of the orientation
of the cylinder of tissue

Fig. 3-7   The flow of auxin through coleoptile tissue is unidirectional.

Another phenomenon worthy of note, at least in some organs, is the polar transport of auxin (1). In an *Avena* coleoptile, for example, if auxin is applied at the apical (A) end it can be collected from the basal (B) end. This directionality of flow from A to B is maintained even if the entire assembly is inverted so that A is below B (Fig. 3-7). If physiological levels of auxin are applied to B, no auxin can be collected at A, irrespective of the position of the entire assembly in space. Although this polarity of auxin transport has been much investigated, relatively little is known about its physiological basis, except that aerobic metabolism and membrane integrity must be maintained.

It used to be thought that auxin was produced in root tips, as well as in stem tips, and that auxin was transported polarly away from root tips toward the stem. Recent evidence (2) implies, however, that the auxin of roots is polarly transported *toward* the apex of the root. Such auxin presumably comes from the stem apex and moves downward in response to the polar transport mechanism. Presumably, the auxin is ultimately destroyed in roots, which are known to be extremely sensitive to auxin and to possess very high levels of the enzyme peroxidase, which can inactivate IAA (3).

When large quantities of auxins, especially of the synthetic type, are applied to intact plants or to excised portions thereof, this polarity of transport may not be so strict. Probably in such instances some of the auxin is transported in the xylem rather than through living cells. Since the xylem tracheids and vessels do not possess membranes, it is difficult to see how they could maintain any directionality of transport. The effect of certain of the synthetic growth regulators, such as 2,3,5-triiodobenzoic acid (TIBA), is probably attributable to its effects on the polar transport mechanism, since a ring of this compound applied around a petiole or stem can be shown to prevent downward auxin movement.

Among the effects produced by exogenously applied auxin is an increased growth of isolated plant parts, such as subapical cylinders cut from oat coleoptile, pea stem or sunflower hypocotyl. When similar concentrations of auxin are applied to intact plants, relatively little effect is produced. This is usually explained by saying that as long as the plant has its normal auxin-synthesizing centers in the apex of the stem or coleoptile, it produces all the auxin required for its own normal growth needs. In such situations, the application of exogenous auxin is without significant additional effect on growth. Yet it is curious that the application of auxins to entire plants does produce effects on the orientation of leaves and sometimes curvature of stems, without inducing significant alteration of the overall growth rate of stems and roots. These facts have led some to the belief that auxin does not really control the rate of elongation under conditions of symmetrical straight growth in intact plants. Instead, because of the polarity of its transport, many prefer to believe it is concerned mainly with tropisms.

When auxin is applied to excised plant parts, increasing the concentration of the auxin increases its effect up to a maximum, after which further increases are inhibitory. The levels that are inhibitory vary, however, from tissue to tissue, roots showing the lowest optimum concentration, stems the highest, and buds intermediate (Fig. 3-8). The response of the plant

Fig. 3-8   Plant organs all show an optimum growth promotion with varying auxin concentrations, higher concentrations being inhibitory. Roots show the lowest optimum concentration, buds intermediate and stems highest. (From Thimann. 1937. Amer. J. Bot. **24**:407–412.)

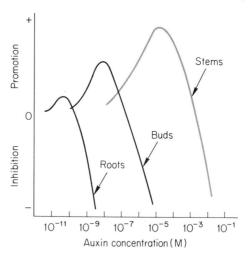

Fig. 3-9   Camellia cuttings. (Left) Control. (Right) Basal end of cutting dipped in a water solution of a mixture of equal amounts of the potassium salts of indolebutyric and naphthaleneacetic acids (total concentration 20 mg/ml). (Photograph courtesy of Boyce Thompson Institute for Plant Research.)

to endogenous or applied auxin is, therefore, dependent both upon the nature of the tissue and the concentration of auxin present.

When auxin is applied to an excised stem, polar transport causes its rapid accumulation at the base of the stem. There, after some time, the accumulated auxin will lead to the production of a swelling or *callus* containing many parenchymatous cells produced either by newly developed meristematic centers, or by activation of existing meristems. Frequently adventitious roots develop in profusion following activation of cells of the cambium (Fig. 3-9). This action of auxins, especially of synthetic analogs such as indolebutyric acid, is so dependable that it is a widely used horticultural device. Cuttings of various plants that do not normally root spontaneously may be dipped basally into solutions or powders of auxin, and then planted in a sand rooting bed. The extensive development of roots follows within days to weeks, depending on the plant.

Another vivid demonstration of the great potential of auxin in controlling cell division can be seen in its action on the cambium of woody plants, especially trees. The cambium ceases division in the autumn, usually in response to decreasing photoperiods, and possibly also in response to the accumulation of the growth inhibitor abscisic acid (see Chapter 6). In the spring, the resumption of cambial activity is at first localized under regions of auxin synthesis, such as developing buds. The artificial injection or application of auxin to such systems stimulates divisions of the cambium to produce xylem elements, whereas with the application of gibberellin, differentiation of phloem tissue from the cambial derivatives occurs (4). In *Ailanthus altissima,* the production first of xylem and later of phloem during the growing season can be correlated with the endogenous levels of auxin and gibberellin (Fig. 3-10).

Auxin also appears to influence the competition between the various buds of a shoot (5). In plants showing extreme apical dominance, only the terminal bud will grow, the other buds being repressed. If the terminal

Fig. 3-10   Changes in (a) stem diameter and (b) hormone levels in *Betula pubescens* and *Ailanthus altissima* following initiation of short-day treatments. The initial xylem production, correlated with auxin content, ceases after 3 weeks. Further stem increase in *Ailanthus* is due to phloem production, which is correlated with gibberellin. [From data of Digby and Wareing (4).]

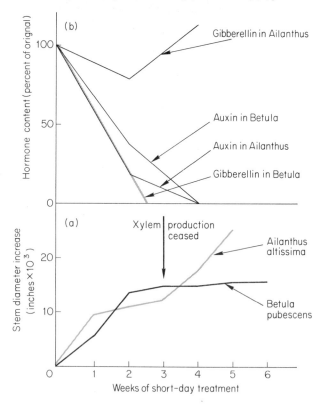

bud is excised, one or several of the buds below it will start to grow, one of these usually becoming dominant. If, following excision of the terminal bud, some auxin is applied to the cut surface, then the lateral buds do not develop. This implies that auxin coming from the apical bud inhibits the development of the lateral buds. There is recent evidence that the inhibition versus growth of lateral buds involves a competitive action of two growth hormones, auxin coming from the apex, and *cytokinins*, (see Chapter 5), which probably come from the root system. Thus, even in a system inhibited by terminally produced auxin, the localized application of cytokinins to a repressed bud may cause it to grow selectively. Once the lateral bud has overcome the inhibition, applications of auxin are no longer inhibitory and may, in fact, enhance lateral stem growth.

The prodigious growth of the ovary into a fruit is another auxin-stimulated phenomenon. Normally, following pollination and double fertilization in the embryo sac, the ovary wall, which ultimately forms the fruit, is stimulated to tremendous growth, through an increase in both cell number and cell size. The role of auxin in this process is indicated by the fact that the auxin levels of ovaries and fruit increase dramatically during such development (Fig. 3-11), and that the application of exogenous

Fig. 3-11   The growth of strawberry receptacles ("fruit") is associated with a rise in the auxin content of the achenes (the true fruit) on the receptacle during the early period of fruit development. (From Nitsch. 1950. Amer. J. Bot. **37**:211–215.)

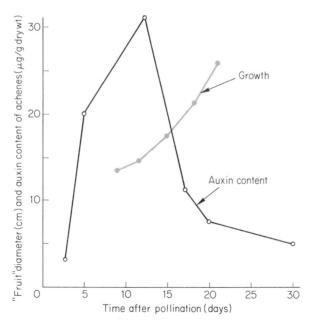

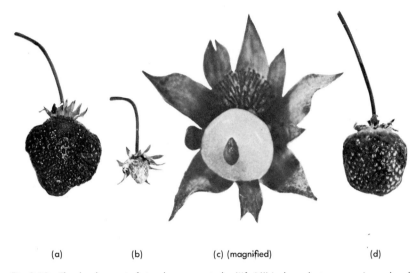

(a)   (b)   (c) (magnified)   (d)

Fig. 3-12   The development of strawberry receptacles ("fruit") is dependent upon auxin produced by the developing achenes. (a) Normal fruit. (b) Fruit of same age with all achenes removed. (c) Fruit with all achenes removed except one. The receptacle develops only under this achene. (d) Fruit with all the achenes removed and replaced by lanolin paste containing 100 ppm β-naphthoxy acetic acid. (From Nitsch. 1950. Amer. J. Bot. **37**:211–215.)

auxins can enhance the process during normal development, even substituting in some respects for normal pollination. The requirement for auxin is a multistage process, auxin being produced following pollination, following fertilization and by the developing seed. In some plants pollination alone produces enough auxin to stimulate fruit development, while in others the complete fertilization or seed development is required (Fig. 3-12). In some fruits auxin alone is optimally effective, while with others it is essential to add gibberellin as well as auxin in order to produce the ovary stimulation.

Auxin-induced ethylene formation

We have described earlier how the sudden production of ethylene plays a dominant role in the ripening of many fruits. Because of the effect of auxin in stimulating the development of the ovary into a fruit, questions have been raised about the possible relation between auxin and ethylene in the control of fruit development. Recently it has become clear that the well-known optimum curve for the effect of auxin on the elongation growth of many isolated plant parts is a consequence of two independent activities of auxin. At the lower concentration levels, i.e., between 10^{-8} and 10^{-6} molar, auxin promotes growth progressively with increasing concentration; at or near the optimum concentration for growth, auxin begins to induce

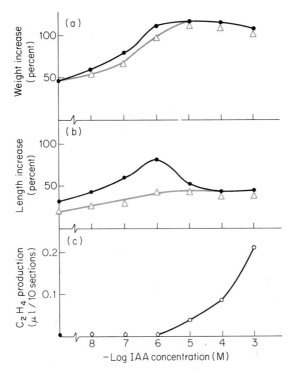

Fig. 3-13 The effect of various IAA concentrations in inducing ethylene formation and its effect on growth in etiolated pea stem sections during 18 hours in sealed flasks. ● = no ethylene; △ = 10 ppm C_2H_4 added. (a) Increase in weight—ethylene has no effect. (b) Increase in length—ethylene is inhibitory, but in the presence of ethylene IAA itself never becomes inhibitory. (c) Ethylene production—this coincides with the optimum growth promotion of IAA in (b). With increasing IAA concentrations the auxin-stimulated ethylene formation results in a growth reduction. [From Burg and Burg (6).]

the formation of ethylene (6), and with successively increasing levels of auxin beyond this point, the tissue is progressively inhibited by the increased levels of ethylene formed. In the presence of external ethylene it has been shown that auxin itself is never inhibitory but only promotes growth (Fig. 3-13). Thus the effect of supraoptimal concentrations of auxin in such plants as sunflower and pea can be ascribed to the ethylene formed. In oat coleoptiles, although high auxin concentrations do produce ethylene, this phenomenon does not correlate well with the growth inhibition. Therefore, we still must find other explanations for some of the inhibitory effects of auxin.

The role of endogenous auxin-induced ethylene production in normal growth regulation is gradually becoming clearer. As previously described, apically applied auxin inhibits the development of subjacent lateral buds, and this action is opposed by cytokinin. It is now evident that the effect of auxin is, in part at least, mediated by ethylene (7), since ethylene production is stimulated by auxin application to buds and the gas inhibits bud growth. Kinetin, which counteracts the action of auxin, also counteracts the effect of applied ethylene.

It has been proposed that the curvature response of pea roots to gravity is due to ethylene production and therefore growth inhibition on the lower sides of the root, which develop supraoptimal auxin concentrations as a result of lateral auxin transport (8). Though auxin applied to pea roots does elicit ethylene production and this gives a growth inhibition, the theory has been criticized on the basis that growth responses in pea roots caused by supraoptimal auxin concentrations are different from those caused by ethylene alone (9). Other mechanisms may be operating here also. A clear case of the promotion of ethylene production by gravitational stimulation is found in the pineapple plant (10). Here, flowering may be induced either by application of synthetic auxin, application of ethylene (11) or simply by prostrating the normally erect stem. In the last technique, the lateral transport of auxin under the influence of the gravitational field seems to cause sufficient accumulation of this substance to elicit ethylene production and the subsequent initiation of floral primordia (Fig. 3-14).

The mode of action of auxin

How can one explain the varied physiological effects produced by auxin? Does the molecule act at only one site? Is the same initial chemical transformation produced by auxin irrespective of the ultimate physiological effect, or must we consider that the auxin molecule is itself versatile, capable of stimulating the functioning of vastly different bio-

Fig. 3-14   When a pineapple that is vegetative (right) is tipped on its side, the dome-shaped apex (seen here in longitudinal section) changes to the initiation of floral primordia, becoming much more pointed (left). This is caused by the production of ethylene resulting from a geostimulated auxin redistribution. (From van Overbeek and Cruzado. 1948. Amer. J. Bot. **35**:410–412.)

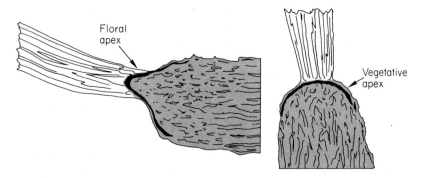

chemical systems? The answers to these questions cannot yet be given, and, in fact, auxin function has been investigated in great detail for only one phenomenon, the enlargement of plant cells.

Auxin and the cell wall

If varying concentrations of auxin are applied to cells capable of elongating, such as those in sections of oat coleoptile, pea epicotyls or sunflower hypocotyls, then increasing doses of auxin produce increasing effects up to an optimal concentration. Usually growth effects can be produced by as little as 10^{-7} M auxin and the optimum is somewhere in the vicinity of 10^{-6} to 10^{-5} M. Concentrations beyond this lead to progressively less promotion and eventually inhibition. As mentioned earlier, the inhibitory phase of this response is probably connected with the promotion of ethylene production. In cases where ethylene production is not the sole cause of the growth inhibition, we do not yet understand how auxin inhibits growth.

Auxin has been shown to produce many physical, chemical and physiological changes within cells, some of them extremely rapidly. For example, auxin is known to promote cytoplasmic streaming in excised oat coleoptile sections within a few minutes after its application (12). Although this rapid response may be the first manifestation of auxin action, it is difficult to relate it directly to cell elongation, which must at some point involve basic changes in the cell wall. Unlike animal cells, plant cell protoplasts are enclosed in a semi-rigid "wooden" box, the cell wall. No matter what changes occur inside the box, the cell cannot enlarge unless the walls can be extended. An analogous situation in animals would be the growth of an arthropod (e.g., lobster), whose entire body is encased in a rigid exoskeleton made largely of chitin. The animal can grow only when the old rigid exoskeleton is shed ("moulting") and the new cuticle underneath is for a time soft and stretchable. Plant cells cannot cast off their cell walls, and the only way they can increase in volume is through extension of the existing cell wall, a process in which auxin has been shown to exert an influence.

If stem or coleoptile tissues are killed by boiling so that cell turgor is removed, their tensile strength and extensibility depend entirely on their cell walls. In such systems it is found that extensibility can be divided into two components, *elasticity* and *plasticity*. Measurements of these separate components may be made by hanging weights on the base of a tissue section and measuring reversible and irreversible extension under load. When the weights are removed the stretched tissue shrinks somewhat but not completely back to its original length (Fig. 3-15). Modern experiments on cell wall extensibility are done using "stress/strain analyzers" which

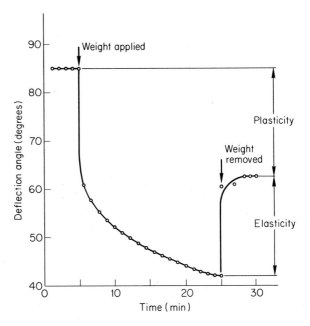

Fig. 3-15 The deflection of oat coleoptiles when a weight is hung on the tip can be used to measure the extension properties of the cell wall. When the weight is removed, some but not all of the deflection is recovered. The reversible deflection is termed *elasticity* and the nonreversible, *plasticity*. More modern methods to measure plastic extensibility use mechanical stress/strain analyzers. (From Tagawa and Bonner. 1957. Plant Physiol. **32:**207–212.)

measure the force required to pull the ends of the tissue apart at a constant slow rate. The permanent deformation is referred to as plastic, while the reversible deformation is called elastic (13). When the tissue is treated with auxin before the extensibility is measured, plasticity is increased (Fig. 3-16).

Cells do not expand in all dimensions under the influence of physiological auxin concentrations; they usually only elongate. Some specific changes must occur within the cell wall to increase the plasticity in the longitudinal direction only. Once wall plasticity has been increased through the action of auxin, the restraining pressure of the wall on the protoplast is decreased. This augments the water potential gradient toward the interior of the cell, since water entry has previously been restrained by wall pressure. Water accordingly diffuses into the cell vacuole, which contains osmotically active solutes. The cell volume is thereby increased and the wall irreversibly stretched.

Let us now examine some of the experimental evidence bearing on auxin action. If tissue sections are placed in a relatively inert isotonic medium such as mannitol, net water entry into cells is prevented, thus making cell expansion impossible. If auxin is added to this medium, there is still no

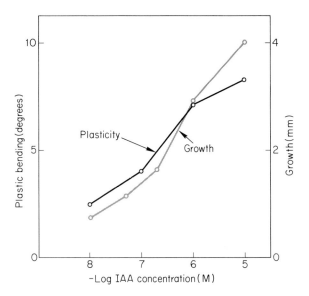

Fig. 3-16   The plasticity of the cell wall as measured in Fig. 3-15 increases in the presence of auxin approximately to a similar extent as the increase in growth. It is, therefore, likely that growth is controlled by the "loosening" of the cell wall under the influence of auxin. (From Bonner. 1960. Z. Schweiz. Forstr. **30**:141–159.)

growth. However, upon removal of the auxin-treated tissue to water, even in the absence of auxin, elongation occurs to the same extent as it would have if the mannitol had not been initially present (Fig. 3-17). From this experiment it can be concluded that auxin is required for some wall loosening activity preparatory to cell elongation, but not for the elongation itself. In the absence of metabolic activity, the primary cell wall is moderately elastic, but no irreversible deformation occurs at normal turgor pressures. The wall structure with which we are concerned cannot, therefore, be changed physically, but requires metabolic reactions. This is borne out by the fact that the auxin-induced wall loosening does not occur in the absence of oxygen and is inhibited by respiratory inhibitors, which prevent the formation of ATP through oxidative phosphorylation. Metabolic activity is, however, not required for the subsequent cell elongation that follows the wall-loosening step.

Auxin and enzymes

In some manner auxin must produce changes in cell wall extensibility. The wall is known to consist predominantly of cellulose microfibrils which are probably substantially crosslinked into a semirigid framework,

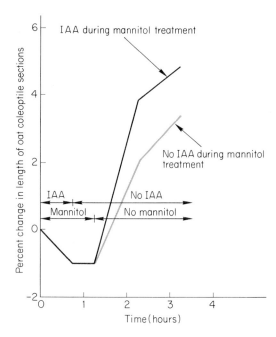

Fig. 3-17 By the use of a hypertonic concentration of mannitol, oat coleoptiles can be osmotically inhibited from expansion, but will expand if removed into water. If auxin is present during the incubation in mannitol, no growth occurs, yet these sections show a greater growth than nonauxin-treated sections when removed to water without auxin. Auxin is considered to cause "wall loosening," which is then followed by wall extension under the influence of osmotic water uptake. This latter process does not require the presence of auxin. (From Cleland and Bonner. 1956. Plant Physiol. **31:350–354.**)

possibly through other glucans. One possibility is that auxin causes the breakage of the crosslinks holding the cellulose microfibrils, permitting movement of one fibril past another and thus wall extension. The breakage of crosslinks is probably enzymatic and the necessary enzymes may be induced to form by auxin. This is the process which may occur during the temperature-dependent lag phase between auxin application and the resulting growth response.

Much evidence accumulated since 1965 supports this hypothesis of auxin action. Auxin promotes the incorporation of precursors into both RNA and protein. In addition, RNA and protein synthesis seems to be required for auxin-induced growth (Fig. 3-18) and the increase in wall plasticity (Fig. 3-19), since both these processes are prevented in the presence of protein and RNA synthesis inhibitors. One of the RNA's synthesized in response to auxin treatment appears to be a messenger RNA essential to the auxin growth response (14, 15).

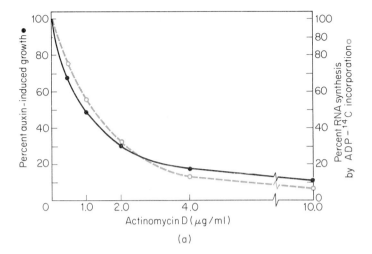

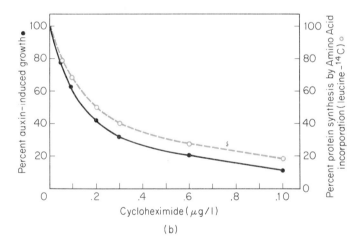

Fig. 3-18 The parallel inhibition of auxin-induced growth and (a) DNA-like (messenger) RNA synthesis by actinomycin D, and (b) protein synthesis by cycloheximide indicate a close correlation between these processes. [From Key, et al. (15).]

Attempts have been made to identify the enzymes involved. Numerous polysaccharide-degrading enzymes such as cellulase have been found to increase in response to auxin treatment but usually over much longer periods than required for the initiation of growth. The increase in wall plasticity is, however, reversed by the addition of metabolic inhibitors

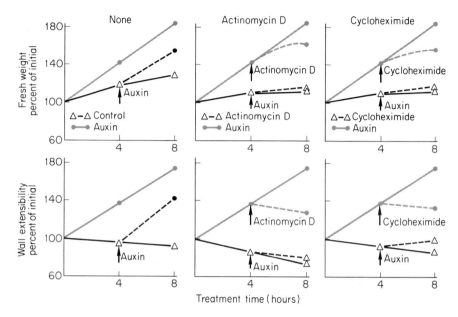

Fig. 3-19   The parallel effects of inhibitors of RNA synthesis (actinomycin D) and protein synthesis (cycloheximide) on wall extensibility and growth of soybean hypocotyls in the presence of auxin provide further evidence that RNA and protein synthesis are required for auxin-induced growth mediated by wall plasticity. (From Coartney, et al. 1967. Plant Physiol. **42**:434–439.)

following the auxin-induced plasticity increases. This makes it unlikely that wall loosening is mediated by polysaccharide-degrading enzymes, since their action is essentially irreversible (16) (Fig. 3-20). Thus, we are faced with the requirement for a reversible enzyme whose equilibrium depends upon whether or not a product of respiration is present. Perhaps the crosslinks between the cellulose microfibrils are other noncellulosic materials on which a reversible enzyme might act. One recent suggestion is that the crosslinks may be hydroxyproline-containing glycopeptides (17) (polysaccharide chains joined by a polypeptide chain containing a high content of the unusual amino acid hydroxyproline). Whether hydroxyproline is or is not present in the cell wall of plant tissue is, however, being argued at the present time, so that it is not now possible to draw any definite conclusion on the nature of the crosslinking.

Both breakage and resynthesis of crosslinks would be required for irreversible extension of the cell wall. Auxin has been shown to promote enzymes involved in wall synthesis and wall synthesis itself, but not without prior cell extension. The new cell wall formation promoted by auxin appears to be deposited within the existing cell wall and to consist largely of hemicellulose.

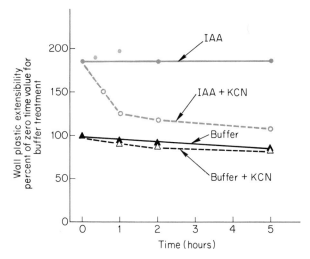

Fig. 3-20   Auxin-induced wall extensibility is reversed in the presence of cyanide. This raises the question of what type of enzyme system must be involved. [From Cleland (16).]

The rapidity of auxin-induced growth

The main difficulty in the proposal that auxin affects growth through enzyme induction is the time required for changes in RNA and protein synthesis. Though a stimulation of RNA synthesis by auxin in 10 minutes has been claimed (18), no protein changes have been detected in less than one hour. Against this is the observation by several workers (19) that growth rate increases occur in 10 minutes or less following auxin application (Fig. 3-21) and the lag between auxin application and the growth response is unchanged by the addition of RNA and protein synthesis inhibitors. By the use of high auxin concentrations and an elevated treatment temperature, which over short periods of time are not supraoptimal, a complete abolition of the lag has now been reported (20) (Fig. 3-22). This clearly rules out gene activation as the primary mode of action of auxin as, even in the most rapid systems known, enzyme production still requires time intervals of several minutes.

Further investigation on the effects of auxin on cell wall properties has also changed our earlier ideas. Auxin was thought to influence only wall plasticity but has now been found to initially increase the elasticity of the cell wall at about the same time as the first increase in the growth rate occurs (under 10 minutes) (21) (Fig. 3-23). The increase in elasticity soon stops, however, and is followed by an increase in the wall plasticity starting about 20 minutes after the application of auxin (22) (Fig. 3-24).

Although the initial mode of action is therefore unknown, it is clear

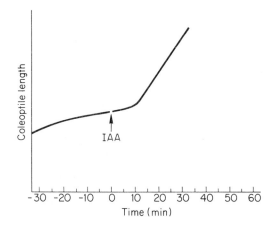

Fig. 3-21   The rate of growth of oat coleoptiles increases sharply 10 minutes after the addition of auxin. [From Evans and Ray (19).]

Fig. 3-22   The lag period between auxin application and the subsequent growth response in oat coleoptiles decreases with temperature so that at 40° the lag period is abolished. This is probably due to the rapid entry of auxin at the high temperature and concentration used. The almost instantaneous response to auxin application rules out an initial action of auxin on protein synthesis. [From Nissl and Zenk (20).]

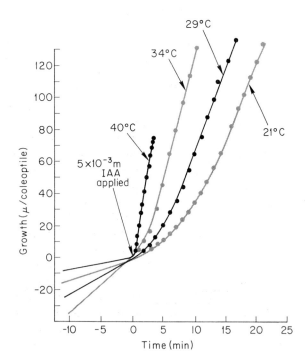

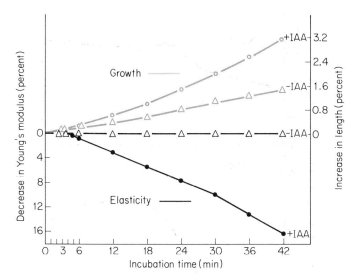

Fig. 3-23   Changes in the length and modulus of elasticity (Young's modulus) of sunflower hypocotyls under the influence of auxin (added at time 0). An increase in elasticity (decrease in the modulus) begins four minutes after the addition of auxin, followed by a growth increase starting two minutes later. [From Uhrström (21).]

that auxin must act on some preformed system, which could be a cellular membrane, a preformed enzyme in a cell membrane or wall, or the wall itself. Such action could result in transitory increases in the elasticity of the cell wall, which in turn would permit an acceleration of the cell growth rate. The new rate cannot, however, be maintained unless wall plasticity is also altered. This could occur slightly later, probably as the result of the action of enzymes, whose synthesis has been stimulated by auxin, acting on the crosslinks in the cell wall. Thus, at the present time, auxin action can best be envisaged as a dual action in the cell wall. Many questions still have to be answered. How does auxin cause the initial growth stimulation? In the following continuation of the growth response, exactly how does auxin promote enzyme formation and what are the enzymes involved? Since even the nature of the crosslinking in the cell wall is not clear, we cannot expect an immediate clear answer.

Structure and activity of auxins

Another approach to deducing the mechanism of action of auxin has been to synthesize large numbers of different auxin analogs, and to attempt to determine the minimum structural requirements for auxin action by a study of their action on cell elongation. Many analogs of

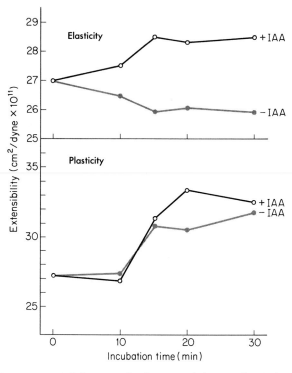

Fig. 3-24   The time course of changes in the elasticity and plasticity of oat coleoptiles after the addition of auxin. Changes in elasticity are first in evidence followed later by plasticity differences. [From Masuda, (22.)]

indoleacetic acid, benzoic acid, phenylacetic acid, naphthylacetic acid and phenoxyacetic acid have been made. A study of their structure-activity relations permits the conjecture that the requirements for activity include: (1) a ring or an easily cyclizable chain, (2) at least one double bond in the ring, (3) a side chain on the ring, adjacent to the double bond, (4) a carboxyl group or potential carboxyl group separated from the double bond by a certain minimum length, generally one or two carbon atoms and (5) correct isomer geometry (Fig. 3-25). Thus, although indoleacetic acid (2C chain) is highly active, indolecarboxylic acid (1C chain) is virtually inactive, and indolebutyric acid (4C chain) is generally less active than IAA [Fig. 3-25 (a)]. With phenoxy compounds, there is a requirement that at least one position *ortho* to the side chain be open. Thus, 2-chlorophenoxyacetic acid is active; the 2,4-dichloro compound is more active, but the 2,6- and 2,4,6-compounds are completely inactive [Fig. 3-25 (b)]. In contrast, in the benzoic acid series it is essential that *both* ortho positions be substituted for activity to occur. Thus, 2-chlorobenzoic acid is inactive,

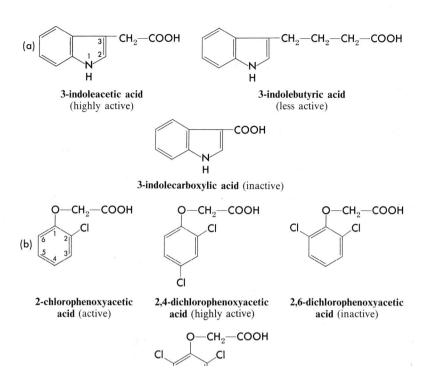

(a)

3-indoleacetic acid
(highly active)

3-indolebutyric acid
(less active)

3-indolecarboxylic acid (inactive)

(b)

2-chlorophenoxyacetic
acid (active)

2,4-dichlorophenoxyacetic
acid (highly active)

2,6-dichlorophenoxyacetic
acid (inactive)

2,4,6-trichlorophenoxyacetic acid (inactive)

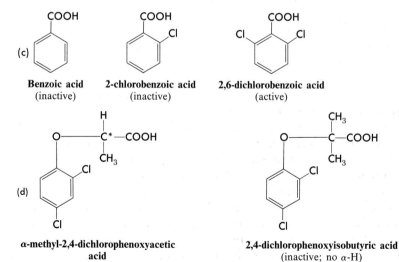

(c)

Benzoic acid
(inactive)

2-chlorobenzoic acid
(inactive)

2,6-dichlorobenzoic acid
(active)

(d)

α-methyl-2,4-dichlorophenoxyacetic
acid

2,4-dichlorophenoxyisobutyric acid
(inactive; no α-H)

(C* is asymmetric; only one
enantiomorph is active)

Fig. 3-25 The structure of various natural and synthetic auxins with relation to their growth-promoting activity.

but the 2,6- and 2,4,6-compounds are completely inactive [Fig. 3-25 (b)]. the substituted phenoxyacetic acid type, replacement of one of the hydrogens on the methylene group by, let us say, a methyl group confers optical activity on the previously symmetrical methylene carbon atom. Where both isomers are available, it is clear that only one is active. If both of the methylene hydrogens are substituted by methyl groups, to yield the phenoxyisobutyric acid, the resulting compound is not only inactive, but serves also as an inhibitor of auxin action [Fig. 3-25 (d)]. It appears that such an auxin analog retains the spatial specificity permitting it to attach to the hypothetical auxin receptor site, but then, by virtue of its own inactivity, prevents other auxin molecules from attaching and acting.

Sometimes, weak auxins are converted to strong auxins and vice versa by action of plant enzymes on the fed compounds. For example, ω-carboxyl phenoxyalkylcarboxylic acids are β-oxidized to yield successively shorter and shorter chains. If the starting number of carbon atoms is even, the product will be the phenoxyacetic acid, a more active auxin. If the starting number of carbon atoms is odd, the final product will be a phenol, which may either be inactive, or may inhibit growth (Fig. 3-26). This phenomenon is used when these substances are employed as herbicides in agriculture. Only certain plants are capable of performing this β-oxidation, so while certain concentrations of 2,4-dichlorophenoxybutyric acid are toxic to a plant that oxidizes the butyric down to the acetic derivative, a plant that is incapable of β-oxidation is unaffected. An example of the latter type of plants is clover, which will remain unharmed while weeds among the clover will be destroyed.

Some workers have attempted to make simple generalizations to explain these varied effects, such as that auxins are molecules that fit into particular niches on membranes and must, therefore, possess a hydrophilic and hydrophobic end with proper spatial dimensions. Others have involved purely geometric considerations, such as a standard distance of 5.5 Å between an electrophilic and nucleophilic site in the molecule (23). Although such generalizations agree with a large number of observations associated with auxin function, no one statement yet satisfactorily accounts for the structural requirements of auxin active molecules.

From the metabolic point of view we know that fed auxin becomes conjugated with various molecules. As of the moment, we have no reason to believe that any of these metabolites is any more closely related to auxin function than is the free acid itself, and, as we shall see, studies of tropisms tell us that it is the free, unattached mobile auxin that is the important correlation carrier in the normal physiology of the plant. Thus, all auxin conjugates so far found may represent sequestration products or detoxification products of the active molecule. It appears that although we know a good deal about auxin metabolism and the physiological effects produced by the administration of auxin, we know essentially nothing

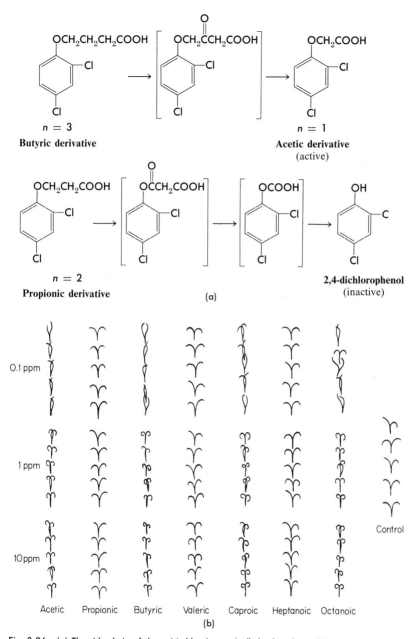

Fig. 3-26 (a) The side chain of the ω-(4-chlorphenoxy) alkylcarboxylic acids is cleaved 2 C's at a time by β oxidation. If the final product possesses one methylene carbon, the product is active, whereas if this is lost so is the activity. (b) Thus, compounds with 2 (acetic), 4 (butyric), 6 (caproic) and 8 (octanoic) carbons in the side chain are active, whereas those with 3 (propionic), 5 (valeric) and 7 (heptanoic) carbons are inactive. The test pictured here is the split pea stem test where the activity is measured by the angle of inward bending of the 2 halves of a split pea stem. With the active compounds at 1 and 10 ppm the half stems have completely crossed and coiled. (From Wain. 1964. Pp. 465–481 in Audus [ed.], The physiology and biochemistry of herbicides. Academic Press, New York.)

concerning the master reaction governing auxin effects in the cell. The same situation holds for almost all hormones in the plant and animal field.

Geotropism

Let us now return to the phenomenon of geotropism and attempt to analyze the physiology of this complex response in terms of what is known about auxin (24). Initially, we can divide the process into three parts:

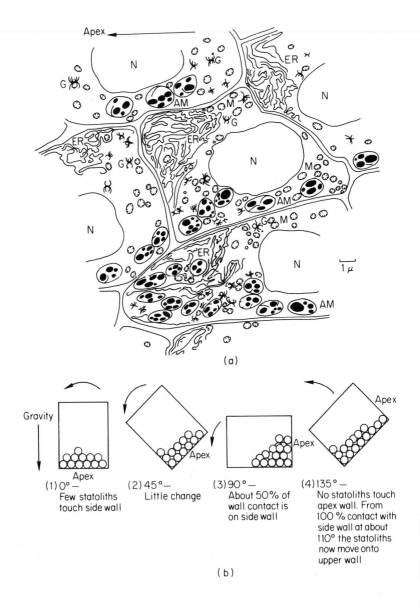

(a)

(1) 0°—
Few statoliths
touch side wall

(2) 45°—
Little change

(3) 90°—
About 50% of
wall contact is
on side wall

(4) 135°—
No statoliths touch
apex wall. From
100% contact with
side wall at about
110° the statoliths
now move onto
upper wall

(b)

1. There must be a sensor for the gravitational stimulus.

2. The sensed stimulus must somehow be amplified.

3. The stimulus must be transduced into differential growth on the two sides of the plant axis, probably as a result of the induced differences in auxin concentration.

It has long been supposed that special starch grains called *statoliths* are active as gravity sensors. One can, in fact, see such grains in geosensitive regions, and can note their displacement as the organ is turned. When the wall on which they have been located is turned to the vertical orientation, they tend to slide down, pile up on and rub against the wall newly located at the bottom of the cell (Fig. 3-27). In the maize root it appears

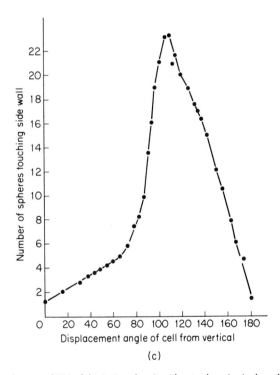

(c)

Fig. 3-27 When the root of *Vicia faba* is tipped on its side, starch grains in the cells of the root cap which previously were situated at the apex end of the cell fall to the new lower side. This can be seen when root cap cells are viewed under the electron microscope (a, opposite) (AM = amyloplasts with enclosed starch grains). The endoplasmic reticulum (ER) also accumulates at the end of the cells towards the root apex but rises when the cell is tipped on its side. No movement of the mitochondria (M), nucleus (N), or Golgi apparatus (G) can be seen. The system can be mimicked by an oil-filled box containing heavier spheres that sink to the lower side as the box is rotated (b, opposite). It is not until the box is rotated past 90 deg to about 100 to 110 deg that a maximum number of spheres touches the lateral wall. This matches very closely the geotropic response to various angles of displacement of plant axes (c). (From Audus. 1962. Symp. Soc. Exp. Biol. **16:**197–226 and Audus. 1964. Physiol. Plant **17:**737–745.)

that all of these sensing cells are in the root cap, for if the cap is removed by delicate surgery, the growth rate of the now naked root tip is unimpaired, but the root is geotropically insensitive until it has regenerated the root cap (25). Similarly, in the rhizoids of *Chara* there are small dense bodies that fall to the floor of the cell and induce curvature. If these bodies are centrifuged out of the cell, or centrifuged into a portion of the cell which is then separated by ligation from the actively growing part of the cell, then there is no tropistic sensitivity, although growth continues. These experiments speak strongly in favor of a sensing function for statoliths, whether they are starch grains in maize roots or chemically undefined particles such as those of *Chara.*

With respect to root geotropism, a special problem is posed. If the auxin of the root comes from the stem, and if the geosensing statoliths are in the root cap, then some mechanism must be found to account for the ability of the root cap to alter auxin distribution in the growing regions of the root itself. Presumably, bioelectric potentials set up following statolith stimulation could be involved.

In opposition to the statolith theory, it has been claimed by several workers that in totally destarched organs, tropistic sensitivity continues (26). Destarching has been accomplished by high temperatures, by acids, and by growth hormones such as gibberellic acid. Certainly the histological data presented in these studies show a great diminution of starch in treated organs, but whether all starch grains have been removed is difficult to say. It is, of course, also possible that starch grains are the normal sensing organelles but that others assume this role if starch is depleted. This would seem a necessary conclusion, since there are some plant organs, such as the aerial roots of the orchid *Laelia,* in which starch does not normally occur, yet these organs are tropistically sensitive.

Geotropically stimulated organs must endure a minimum "presentation time" to the gravitational stimulus if curvature is to occur (Fig. 3-28). In most instances several minutes suffice, although some organs require longer periods. The gravitational field can be substituted or annulled by centrifugal fields. Thus, if germinating seeds are placed on a clinostat revolving at such a rate as to give a force of 1 g at right angles to the gravitational force of the earth, then the roots will grow at an angle of 45 deg from the vertical. In fact, they seem quite competent to solve the force-vectorial problem posed by any combination of centrifugal force and gravity. This response to centrifugal force strengthens the belief that the initial sensing occurs by virtue of the displacement of dense bodies, such as statoliths.

Within minutes after completion of the presentation time there develops across the stimulated organ a transverse electric potential of up to 100 millivolts, the stimulated side being electropositive to the unstimulated side. This was originally thought to be the motive force for the ultimate

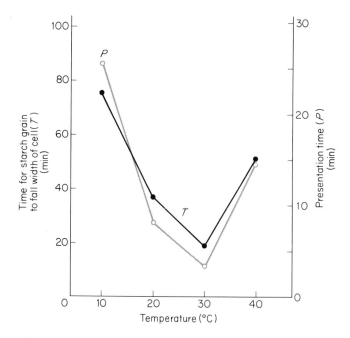

Fig. 3-28   The presentation time, or the time between stimulus and response, in *Lathyrus odoratus* stems varies with temperature, as does the rate of fall of starch grains in the cell due to the changing viscosity of the cytoplasm. The close correspondence of these two parameters is indicative of starch grains being the mechanism by which gravity is perceived. (From Audus. 1962. Symp. Soc. Exp. Biol. **16**:197–226.)

displacement of auxin, since the negative indoleacetate ion would be expected to move electrophoretically to the positive pole. However, it has now been clearly demonstrated that auxin is required for the development of the bioelectric potential and that auxin displacement probably precedes the development of the potential (Fig. 3-29). Measurement of the size of the potential and its electrical sign is a complex process because of the tendency of electrodes to become polarized. Recently, this problem has been satisfactorily overcome by the use of rapidly flowing fluid junctions in contact with the stimulated organs, and better yet by the use of the vibrating reed electrometer, which does not ever come into contact with the stimulated organ.

The occurrence of induced lateral migration of auxin under the influence of a gravitational stimulus has been most clearly demonstrated by the application of labeled indoleacetic acid to treated tissue. It is clear from the work of several investigators that symmetrical application of labeled IAA results in symmetrical distribution of label in a geotropically unstimulated organ, but in the accumulation of label on the lower side

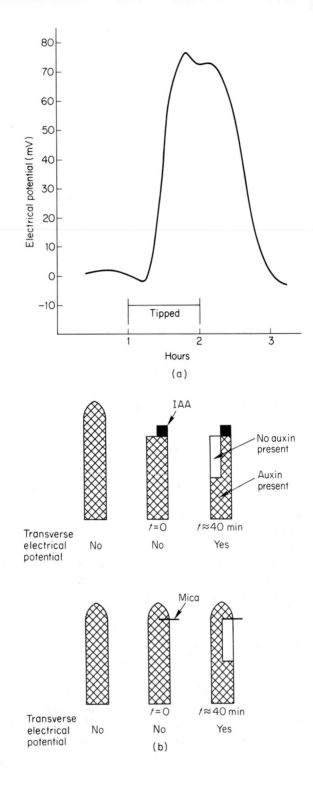

Electrical potential (mV)

Tipped

Hours

(a)

IAA

No auxin
present

Auxin
present

$t=0$ $t\approx40$ min

Transverse
electrical   No   No   Yes
potential

Mica

$t=0$ $t\approx40$ min

Transverse
electrical   No   No   Yes
potential

(b)

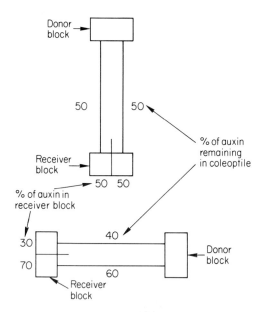

Fig. 3-30 In corn coleoptiles geotropic stimulation produces a redistribution of symmetrically applied IAA. As a result of the higher concentrations on the lower side the coleoptile grows more rapidly on this side and thus turns upward. (Data from Gillespie and Thimann. 1963. Plant Physiol. **38:**214–225.)

of a geotropically stimulated organ (Fig. 3-30). Whether this occurs by a directed lateral migration or by an initial blockade of longitudinal transport causing a pileup and "spillover" of auxin is unknown. A further study of geotropism by physicists, chemists and biologists is obviously required.

Phototropism

In phototropism, as in geotropism, a unilateral stimulus is amplified and transduced into an unequal distribution of auxin (27). By the use of interference filters or a monochromator it can be shown that short-

Fig. 3-29 After a corn coleoptile is tipped on its side it develops a transverse electric potential, but only after a long period (a). That this geoelectric potential only occurs after and not before auxin redistribution is shown by the fact that in vertical coleoptiles an electric potential can be caused by the artificial development of an auxin gradient. This is achieved by the unilateral application of IAA or by a piece of mica (b). These acts themselves have no effect (*t* = 0), but after downward transport has removed the auxin from one part of the coleoptile (*t* = 40 min) an electric potential can be found. (From Grahm. 1964. Physiol. Plant. **17:**231–261.)

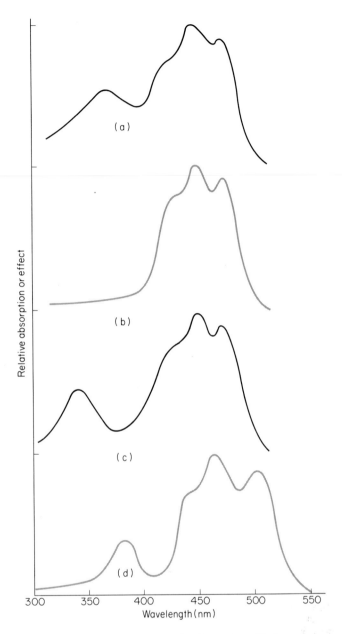

Fig. 3-31   Comparison of (a) the action spectrum for phototropism with absorption spectra of (b) *trans*-β carotene, (c) *cis*-β carotene, unknown in nature, and (d) riboflavin in castor oil. (From Galston. 1967. Amer. Sci. **55**:144–160.)

wavelength visible light is most effective in inducing this process. If dose-response curves are made and the relative efficiency per quantum is calculated, then the action spectrum curves so produced indicate peaks at roughly 480, 440, 420 and 370 nm with a cut-off point of about 500 nm [Fig. 3-31 (a)]. There are scattered reports in the literature of the phototropic efficacy of red light, but these seem not to be part of the same photoreaction that governs usual phototropism. As previously noted, light to be effective must be absorbed, so that the action spectrum should give some notion of the absorption spectrum of the effective pigment. Two classes of pigments, the carotenoids and the flavins, are likely candidates for this role. Carotenoids such as β-carotene have an absorption spectrum that matches the three visible peaks very nicely, but this compound lacks entirely the peak at 370 nm shown in the action spectrum [Fig. 3-31 (b)]. Certain 9-*cis* carotenoids, not known to occur in nature, do possess a peak here [Fig. 3-31 (c)], and there is the possibility that an active compound such as this is isomerized upon extraction. Flavins in lipoidal media also show the peaks in the visible very nicely, and also have a good ultraviolet peak roughly matching that shown for the action spectrum [Fig. 3-31(d)]. On the basis of available data, one cannot decide clearly between these two classes of possible photoreceptors. The only way to settle this quandary is to obtain organisms that are deficient in one or another of the pigment types, and to study the relationship of the induced pigment deficiency to phototropic sensitivity. Through the use of inhibitors of carotenoid biosynthesis it has been possible to produce organisms, containing 20% or less of the normal carotenoid content, which retain normal phototropic sensitivity (28). Attempts to reduce the flavin concentration in the same way have not yet been successful, but the results with carotene depletion permit us to say either that the great bulk of carotene present is not involved in phototropism and only a quantitatively small constituent is involved as receptor, or that carotenes are not the photoreceptors at all. This problem awaits further study.

The rest of the physiology of this tropism resembles that already delineated for geotropism. A transverse electric potential appears sometime after unilateral irradiation. As in geotropism, this potential seems to be dependent upon prior redistribution of auxin. The problems involved in examining the nature of the origin of the transverse potential and its relation to auxin redistribution are unsolved in phototropism as they are in geotropism.

Thigmotropism

Many plant organs are sensitive to mechanical stimuli, and frequently will bend in the direction from which the stimulus is applied. The curvature of a plant organ toward or away from a point of mechanical stimulation is referred to as *thigmotropism*. Some plant organs are sensitive to touch, but since the direction of their movement is independent of the direction from which the stimulus comes, such organs are more properly referred to as *thigmonastic*.

The physiology of thigmonasty has recently been investigated in pea tendrils (29). The area of maximum sensitivity is just behind the tip of the tendril on the concave side. Within several minutes of the initial stimulation there is an unequal contraction, mainly on the ventral side, followed by a differential elongation, which is greater on the dorsal side. Reaction to a single touch stimulus is greatest at about 30 minutes, and is followed by recovery after 1 hour. Repeated stimulation by twining of the tendril around a solid support leads to continued unequal dorsiventral elongation, i.e., coiling.

The coiling of pea tendrils is associated with a rapid drop in the level of ATP in the organ, and a rise in inorganic phosphate. Therefore, it is concluded that an ATPase is activated by touch. Such an ATPase can be demonstrated both in vivo and in vitro and it appears that the ATPase activity decreases following stimulation and increases again during recovery from stimulation. It has been suggested that the ATPase is membrane-bound and acts as an ion pump, affecting water uptake and cell elongation through control of ion movement.

The touch stimulus can be completely supplanted in vitro or in vivo by the symmetrical application of auxin. It is not yet known whether this leads to an asymmetrical distribution of auxin on dorsal and ventral sides, or whether, in fact, the touch stimulus is in some way connected with a lateral displacement of auxin, as in the case of gravitational and light stimuli. It is reasonable to suspect, however, that auxin physiology is somehow involved.

Other tropisms have been reported in the literature. Responses to electricity (electrotropism), magnetism (magnetotropism) and water (hydrotropism) have been described, but since practically nothing is known of the physiology of these responses, they will not be further discussed here.

Circumnutation

As a plant stem grows, its apex does not rise smoothly in a straight line. Rather, growth occurs by a series of jerky movements, in which the apex moves from side to side. If one looks down on the apex from above and traces the pattern of the apex on a horizontal plane, one obtains a figure more or less approximating Brownian movement, with random oscillations around a central point. This complex of movements, referred to as *circumnutation*, is the result of the unequal rates of elongation of the two sides of a plant organ. If a plant organ describing such movements is exposed to a unilateral stimulus, then instead of randomly oscillating about the center, it begins to describe asymmetric excursions from the center, leading to a net component of movement in one direction. It thus appears that all plant tropisms are the results of alterations of a basic pattern of circumnutational movements (30).

These movements are known to be endogenously rhythmic with periods varying from minutes to hours. Morphogenetically active light decreases the period and increases the amplitude of such rhythms. Neither the time-keeping mechanism nor the mode of its resetting is known.

Summary

A growth-promoting material, auxin, is produced at stem tips and moves polarly away from the locale of its production, enhancing the rate of elongation of cells at some distance from the tip. Normally this substance is equally distributed on the two sides of an organ, but where such an organ is stimulated unilaterally by light, gravity or touch, auxin becomes unequally distributed, resulting in unequal growth promotion. Indole-3-acetic acid is the major auxin in many plants.

Among other auxin effects are the production of adventitious roots on stem cuttings, promotion of cambial activity, inhibition of lateral buds and the promotion of fruit growth. The inhibitory effects of high auxin concentrations are the result, at least in some cases, of auxin-stimulated ethylene production. The mode of action of auxin involves the induced synthesis of proteins, which may act as enzymes, allowing extension of the cell wall. Some rapid growth responses of auxin may involve other modes of action.

Gravity is probably perceived by falling statoliths and light by a carotenoid or flavin pigment, which then results in an asymmetrical auxin distribution and a transverse electric potential. The reception of touch, at least in tendrils, involves changes in cell permeability, possibly mediated by membrane-localized ATPase.

REFERENCES

General

Audus, L. J. 1959. Plant growth substances. Leonard Hill, London. 553 p.

Galston, A. W. and P. J. Davies. 1969. Hormonal regulation in higher plants. Science **163**:1288–1297.

Key, J. L. 1969. Hormones and nucleic acid metabolism. Ann. Rev. Plant Physiol. **20**:449–474.

Went, F. W. and K. V. Thimann. 1937. Phytohormones. Macmillan, New York. 294 p.

References 24, 27 and 29 (below) may be used as general references for the various tropisms.

1. Goldsmith, M. H. M. 1968. The transport of auxin. Ann. Rev. Plant Physiol. **19:**347–360.

2. Wilkins, M. B. and T. K. Scott. 1968. Auxin transport in roots. Nature **219:**1388–1389.

3. Galston, A. W. and L. Y. Dalberg. 1954. The adaptive formation and physiological significance of indoleacetic acid oxidase. Amer. J. Bot. **41:**373–380.

4. Digby, J. and P. F. Wareing. 1966. The effect of applied growth hormones on cambial division and the differentiation of the cambial derivatives. Ann. Bot. **30:**539–548.

5. Sachs, T. and K. V. Thimann. 1967. The role of auxins and cytokinins in the release of buds from dominance. Amer. J. Bot. **54:**136–144.

6. Burg, S. P. and E. A. Burg. 1966. Interaction between auxin and ethylene and its role in plant growth. Proc. Nat. Acad. Sci. U.S. **55:**262–269.

7. Burg, S. P. and E. A. Burg. 1968. Ethylene formation in pea seedlings; its relation to the inhibition of bud growth caused by indole-3-acetic acid. Plant Physiol. **43:**1069–1074.

8. Chadwick, A. V. and S. P. Burg. 1967. An explanation of the inhibition of root growth caused by indole-3-acetic acid. Plant Physiol. **42:**415–420.

9. Andreae, W. A., M. A. Venis, F. Jursic and T. Dumas. 1968. Does ethylene mediate root growth inhibition by indole-3-acetic acid? Plant Phys. **43:**1375–1379.

10. van Overbeek, J. and H. J. Cruzado. 1948. Flower formation in the pineapple plant by geotropic stimulation. Amer. J. Bot. **35:**410–412.

11. Burg, S. P. and E. A. Burg. 1966. Auxin induced ethylene formation: its relation to flowering in the pineapple. Science **152:**1269.

12. Thimann, K. V. and B. M. Sweeney. 1937. The effect of auxins on protoplasmic streaming. J. Gen. Physiol. **21:**123–135.

13. Lockhart, J. A. 1965. Cell extension, p. 826–849. *In* J. Bonner and J. E. Varner [eds.] Plant biochemistry. Academic Press, New York.

14. Trewavas, A. J. 1968. Effect of IAA on RNA and protein synthesis. Arch. Biochem. Biophys. **123:**324–335. _____ 1968. Phytochemistry **7:**673–681.

15. Key, J. L., N. M. Barnett and C. Y. Lin. 1967. RNA and protein biosynthesis and the regulation of cell elongation by auxin. Ann. N.Y. Acad. Sci. **144**:49–62.

16. Cleland, R. 1968. Auxin and wall extensibility: Reversibility of auxin induced wall loosening process. Science **160**:192–193.

17. Lamport, D. T. A. 1965. The protein component of primary cell walls. Adv. Bot. Res. **2**:151–218.

18. Masuda, Y. and S. Kamisaka. 1969. Rapid stimulation of RNA biosynthesis by auxin. Plant & Cell Physiol. **10**:79–86.

19. Evans, M. L. and P. M. Ray. 1969. Timing of the auxin response in coleoptiles and its implications regarding auxin action. J. Gen. Physiol. **53**:1–20.

20. Nissl, D. and M. H. Zenk. 1969. Evidence against induction of protein synthesis during auxin induced initial elongation of Avena coleoptiles. Planta **89**:323–341.

21. Uhrström, I. 1969. The time effect of auxin and calcium on growth and elastic modulus in hypocotyls. Physiol. Plant. **22**:271–287.

22. Masuda, Y. 1969. Auxin induced cell expansion in relation to cell wall extensibility. Plant Cell Physiol. **10**:1–9.

23. Porter, W. L. and K. V. Thimann. 1965. Molecular requirements for auxin action. I. Halogenated indoles and indoleacetic acid. Phytochem. **4**:229–243.

24. Wilkins, M. B. 1966. Geotropism. Ann. Rev. Plant Physiol. **17**:379–408.

25. Juniper, B. E., S. Groves, B. Landau-Schachar and L. J. Audus. 1966. Root cap and the perception of gravity. Nature **209**:93–94.

26. Pickard, B. G. and K. V. Thimann. 1966. Geotropic response of wheat coleoptiles in absence of amyloplast starch. J. Gen. Physiol. **49**:1065–1086.

27. Briggs, W. R. 1964. Phototropism in higher plants, p. 223–271. *In* A. C. Giese [ed.] Photophysiology I. Academic Press, New York.

28. Bara, M. and A. W. Galston. 1968. Experimental modification of pigment content and phototropic sensitivity in excised Avena coleoptiles. Physiol. Plant. **21**:109–118.

29. Jaffe, M. J. and A. W. Galston. 1968. The physiology of tendrils. Ann. Rev. Plant Physiol. **19**:417–434.

30. Darwin, C. 1897. The power of movement in plants. Appleton, New York. 592p.

FOUR

Gibberellins

Internal regulatory mechanisms can be uncovered either by an analysis of normal physiological processes (the analysis of phototropism led to the discovery of auxin) or by observation of abnormal physiology (the examination of the harmful effects of illuminating gas on plants was one of the leads to the discovery of the regulatory role of ethylene). The gibberellins, now recognized as having great importance in the normal physiology of the plant, were discovered through an analysis of a plant disease, the so-called bakanae or "foolish-seedling" disease of rice in the Orient.

In the early part of the twentieth century, rice farmers noted that certain of their seedlings grew at an amazingly rapid rate, far outdistancing the other plants in the field. This early advantage came to naught, however, since the seedlings never reached maturity in the sense that they rarely flowered and never produced viable seeds. Examination of these abnormal plants revealed that they were infected by a fungus whose asexual (imperfect) stage was referred to as *Fusarium*. Spores of the fungus could transmit the disease from plant to plant. Later the fungus was identified as an Ascomycete, and was renamed *Gibberella fujikuroi*. When this fungus was grown in culture and a cell-free filtrate of its culture medium was applied externally to uninfected plants, the treated plants developed the hyperelongation symptoms characteristic of the infected plant. The substance producing this effect was named *gibberellin*.

Research on this interesting phenomenon proceeded exclusively in Japan prior to World War II, and the results were reported in Japanese journals.

96

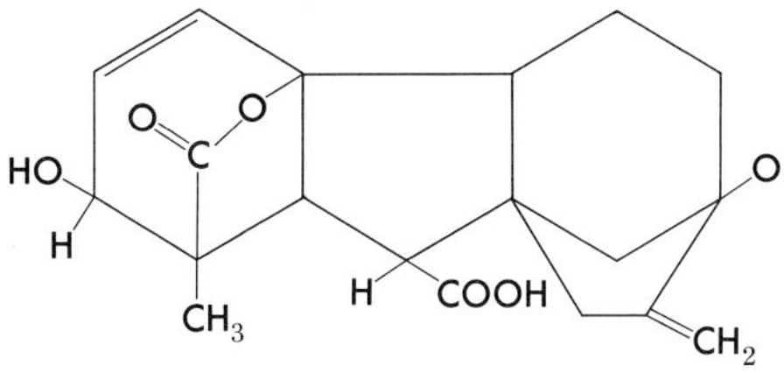

Fig. 4-1   The structural formula of gibberellic acid (GA$_3$).

In the decade following the end of the war, the research was taken up in Britain and America, as well as in Japan, and ultimately led to the isolation and chemical identification of the effective compound. The first active compound to be purified and structurally identified was the substance produced by *Gibberella;* it was termed *gibberellic acid* or GA$_3$. Later, related compounds were isolated; more than twenty different gibberellins are now known, from a variety of natural sources, both fungal and higher plant.

Gibberellins have a fairly complicated structure (Fig. 4-1), chemically isoprenoid in nature. The differences between the various gibberellins reside in the number and placement of double bonds and hydroxyl groups. These various gibberellins display a varied spectrum of biological activity, some plants responding much better to particular gibberellins than to others. Different developmental stages in the same plant may also show different sensitivity to specific gibberellins. In addition to the gibberellins themselves, certain related compounds possessing only a portion of the total molecular structure are also active. One such material is helminthosporol, derived from the fungus *Helminthosporium* (Fig. 4-2).

Following the isolation of the various kinds of gibberellins, plant physiologists began to apply them to many types of plants. It soon became clear that some plants were caused to hyperelongate by external applications of gibberellin, whereas others were not. Soon an important gen-

Fig. 4-2   The structural formula of helminthosporol.

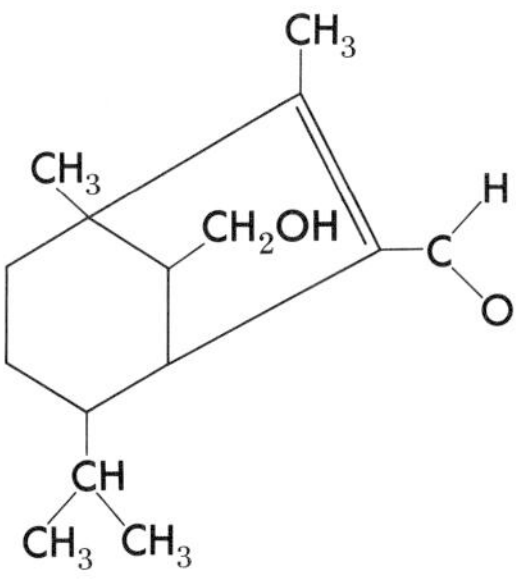

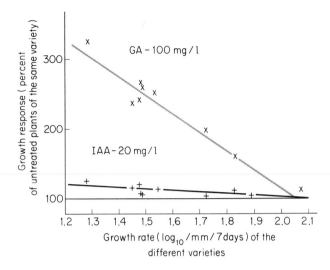

Fig. 4-3   When gibberellin was applied to 11 different varieties of pea plants ranging from dwarf (low growth rate) to tall (high growth rate), their response to the gibberellin was inversely proportional to their growth rate. Auxin had virtually no effect on the growth of the intact plants. (From Brian and Hemming. 1955. Physiol. Plant. **8:**669–681.)

eralization emerged: genetic dwarfs were the most sensitive to gibberellin, whereas genetically tall plants responded only slightly (Fig. 4-3). It appeared that certain dwarfs, such as in maize (Fig. 4-4), in which the stunted habit is known to be due to a single recessive gene, failed to grow tall because they lack the genetic potential for gibberellin production. The exogenous application of gibberellin, therefore, produced a phenocopy of the normal, genetically tall strain. Some tall plants have been found to contain more gibberellin than dwarf plants, although not in all cases. However, when separated chromatographically, the gibberellins of dwarf and tall plants show some qualitative differences, so we may be dealing with a situation in which the gibberellin type differs with genotype, that present in the dwarfs being less active in growth promotion.

Assay for gibberellin

The response of dwarf plants of both maize and pea to external gibberellin furnishes a convenient tool for the assay of these substances in a variety of natural sources. Extracts of gibberellins can be made by organic solvents such as ethyl acetate; these extracts can then be partitioned amongst solvents in such a way as to remove interfering impurities and the now purified and concentrated active principle can be applied

Fig. 4-4   The effect of gibberellic acid on normal and dwarf corn. Left to right, normal control, normal plus gibberellin, dwarf control, dwarf plus gibberellin. (From Phinney and West. 1960. Gibberellins in the growth of flowering plants. *In* D. Rudnick [ed.] Developing cell systems and their control. Ronald Press, New York.)

in various dilutions to dwarf plants. The elongation of the leaf sheath of maize or of the stem of dwarf peas furnishes a convenient index of activity. When referred to the growth produced by known concentrations of gibberellin, the activity of an extract can then be expressed in absolute terms, such as microgram equivalents of gibberellin per milliliter (Fig. 4-5).

The different gibberellins can be separated by chromatographic techniques and the amount of gibberellin present can be assayed by performing a bioassay on the various fractions. For example, in paper chromatography the paper can be cut into portions, the contents eluted and then applied to a test plant. Intensity of color reactions following a spraying of the chromatograms with specific reagents may also be used as an index of concentration, but these assays are much less sensitive and specific than the bioassay and have not been widely used.

By the use of the bioassay, gibberellin has been found most abundant in seeds, especially in the immature "milk" stage. It is difficult to detect gibberellins in mature seeds in such high quantities, although some successful extractions have been made. This presence of gibberellin in seeds naturally led many to inquire into the possibility of some regulatory role of gibberellin in germination. Numerous experiments appeared to substantiate the hypothesis that there is a connection between the two. For example, the seeds of many cereals, such as barley, contain reserve starch

Fig. 4-5   The increased height of intact dwarf peas in response to applications of gibberellin solution can be used to assay the quantity of gibberellin applied. The photo shows dwarf peas 26 days after the application of the gibberellin solution. From right to left, untreated, alcohol control, 0.01 μg GA, then increasing GA doses in twofold steps up to 10.24 μg GA. (From Brian and Hemming. 1955. Physiol. Plant. **8**:669–681.)

which is rapidly hydrolyzed at the onset of germination. If barley seeds containing embryos are soaked, starch hydrolysis begins rapidly. If the embryos are removed prior to soaking, no starch hydrolysis occurs in the embryonectomized seeds. If, now, gibberellin is supplied to such embryoless seeds, starch hydrolysis proceeds (Fig. 4-6). It thus appears that the embryo, shortly after imbibition of the seeds, normally produces gibberellins that activate the starch hydrolysis process. As we shall see later, there is a reason to believe that gibberellin, acting on the aleurone layer, triggers the *de novo* production of specific RNA molecules that code for the starch-digesting enzyme α-amylase. This response, too, can be used as a bioassay for gibberellin in plant extracts (1). Suitably purified extracts can be supplied to embryonectomized barley seeds, and the amount of starch hydrolysis can be determined chemically by analysis of the amount of reducing sugar produced. Again, by reference to a standard curve, gibberellin in the extract can be estimated (Fig. 4-7).

Fig. 4-6   Three sterile halves of barley grain without embryo. To the open surface of each was added (right to left) either 0.5 microliter of water, gibberellic acid at a concentration of 1 part per billion, or gibberellic acid at a concentration of 100 parts per billion. The photograph, taken 48 hours later, shows that digestion of the starch-filled storage tissue is already taking place in the grains treated with gibberellin. The hormone gibberellin promotes production and secretion of the enzymes that cause hydrolysis of the storage material. (Photograph courtesy of J. E. Varner, Michigan State University.)

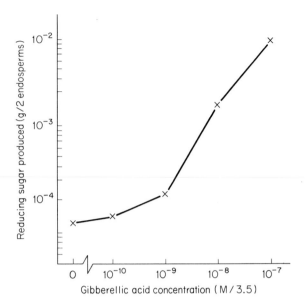

Fig. 4-7   The gibberellin-promoted production of sugars from starch in barley grains seen in Fig. 4-6 can be used as a bioassay of the quantity of gibberellin present. The results shown above are the amounts of sugar produced when 2 half barley grains were incubated in 1 ml of gibberellic acid solution for 36 hours at 30°C. [From Coombe, et al. (1).]

Physiological role of gibberellin in the plant

We have already called attention to the fact that gibberellin produces extraordinary elongation of the stems and leaf sheaths of plants genetically deficient in gibberellin. You will recall that auxin also induces cellular elongation. Is the action of these two hormones the same? Clearly, there are some quite important differences. In the first place, there are very few instances in which the application of auxin to intact plants leads to any increase in elongation at all. Yet it is here that gibberellin produces its most dramatic effect. By contrast, sections of oat coleoptiles, pea epicotyls and sunflower hypocotyls are stimulated to extremely rapid growth by exogenous auxins, and here gibberellin is generally relatively innocuous. In pea epicotyls, it appears that the cells very close to the apical hook respond best to gibberellin and little to auxin, while cells further down the epicotyl respond most to auxin and less to gibberellin. In some systems gibberellin and auxin interact synergistically (Fig. 4-8) (2) and when this occurs it appears that gibberellin action must precede that of auxin. For example, pieces of Jerusalem artichoke tuber, not normally responsive to

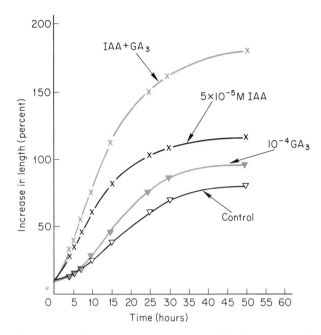

Fig. 4-8   In the elongation of green pea stem sections IAA and GA show a synergistic interaction, producing more growth together than when either is used alone. [From Ockerse and Galston (2).]

auxin, are made very responsive by pretreatment with gibberellin. Green pea stem sections pretreated with gibberellin and later supplied with auxin grow much more than if supplied with either hormone alone, and also more than the sum of the growth increments induced by either hormone alone. Yet, if auxin is given first, followed by gibberellin, growth is much less and it is difficult to achieve even additive effects of the two hormones.

It has been suggested by some workers that gibberellin acts by promoting the synthesis of auxin from its precursor tryptophan or tryptamine. Although analytical data clearly support the notion that gibberellin does increase auxin production, this fact alone cannot explain the various actions of gibberellin, especially those on reproduction. It is probable, rather, that gibberellin activates a multiplicity of biochemical processes, including the conversion of tryptophan to auxin.

In many long-day plants, the photoperiodic stimulus to bolting and flowering can be completely replaced by gibberellin. The effect of gibberellin in the promotion of bolting is to increase the number of cell divisions as well as to contribute to the elongation of the cells produced by such divisions (Fig. 4-9). Where only small amounts of gibberellin are applied, bolting characteristic of the formation of the seed stalk in long-day

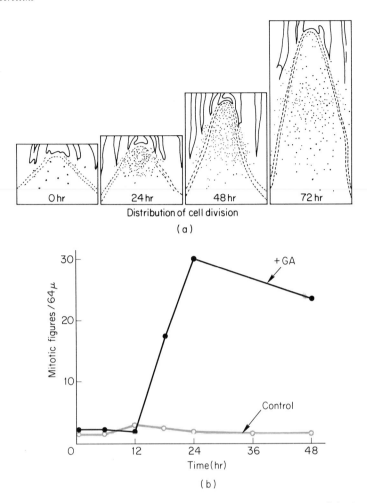

Distribution of cell division

(a)

(b)

Fig. 4-9   Gibberellin applications to rosette plants produce bolting by increasing cell division. This can readily be seen in longitudinal sections through the axis of *Samolus parviflorus* in which division increases following application of GA (a). (Each dot represents one mitotic figure in a 64-micron-thick slice.) The number of such mitoses with and without GA in stem apices of *Hyoscyamus niger* is shown in (b). By the use of AMO 1618, an antigibberellin growth retarder, the meristematic activity in stem tips of *Chrysanthemum moriflorum* can be seen to be directly due to GA (c, opposite). (The heavy lines show the vascular boundaries within which each dot represents one mitotic figure in a 60-micron-thick slice.) (From Sachs. 1965. Ann. Rev. Plant Physiol. **16**:73–96 and Sachs, et al. 1959. Amer. J. Bot. **46**:376–384.)

plants occurs, but floral primordia are not differentiated. Higher quantities of gibberellin generally produce not only bolting, but flowering as well. There are some analytical data to support the contention that in long-day

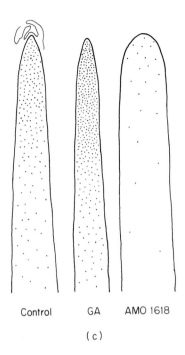

Control   GA   AMO 1618

(c)

plants the administration of a long-day inductive treatment stimulates gibberellin production, which then causes the morphogenetic responses. In other instances gibberellin activity in apices of plants prior to bolting fell as bolting proceeded, indicating that gibberellin may be used up during bolting. Gibberellin is ineffective in causing flowering in short-day plants, and, in fact, appears to work in the opposite direction. In some short-day plants, a related isoprenoid compound called *abscisic acid* (Chapter 6) may act as the analogous stimulatory material.

A study of the responses of plants to applied gibberellin indicates that it produces effects normally controlled by phytochrome or induced by chilling. For example, gibberellin will substitute for the red light promotion of germination in light-sensitive lettuce seeds and the vernalization requirement for flowering in carrot (see Chapter 1). There is evidence that the plant responses to phytochrome and low temperature are, at least in part, mediated through the induction of gibberellin synthesis. Formation of gibberellins, or activation of previously inactive forms, has been found to occur in the response of seeds to both low temperature and red light and in the breaking of bud dormancy by these agents. We shall discuss the hormonal interrelationships controlling dormancy and germination in Chapter 6.

Gibberellin may also act to induce the formation of parthenocarpic

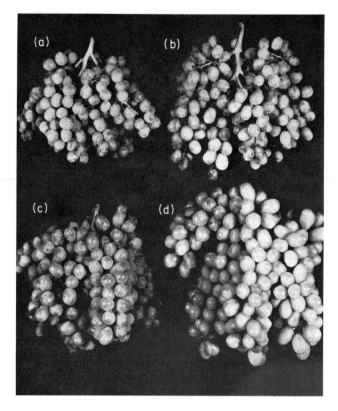

Fig. 4-10   The effect of gibberellin on the growth of Thompson seedless grapes. (a) Control grapes (b, c, d) grapes sprayed with 5, 20 and 50 ppm GA, respectively. (Courtesy of R. J. Weaver and S. B. McCune.)

fruits, either alone or in combination with auxin (Fig. 4-10). An example of this is the apple, whose parthenocarpic development had long been sought and never achieved by the use of auxin alone. A rise in the natural gibberellin content in the seed is also correlated with the maximum period of seed growth (Fig. 4-11), indicating probable gibberellin control of seed development as well.

The mode of action of gibberellin

With gibberellin, as with auxin, we have the problem of explaining how very minute quantities of such a substance can control numerous and varied morphogenetic responses, such as seed germination, cell division, cell elongation and floral initiation. Only one phenomenon has been analyzed in detail, i.e., the induction of starch hydrolysis in embryonecto-mized barley seeds.

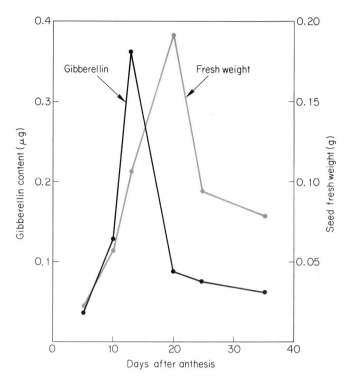

Fig. 4-11 The maximum gibberellin content of developing seeds of *Pharbitis* is correlated with the period of seed growth. (From Murakami. 1961. Bot. Mag. [Tokyo] **74**:241–247.)

Recently, it has become clear that control of starch digestion by gibberellin has to do with the regulation of enzyme production. The application of gibberellin to embryonectomized barley seeds results in the appearance of amylase activity, which in turn causes the hydrolysis of the starch contained in the endosperm of the barley grain (Figs. 4-12 and 4-13). By the removal of the aleurone layer, which is a layer of cells surrounding the storage cells of the endosperm, one can show that the new protein formation occurs in the aleurone (Fig. 4-14). The aleurone, therefore, has the function of producing and secreting hydrolytic enzymes for digesting the food reserves in the endosperm, and it is these aleurone cells that are the "target cells" that respond to gibberellin. This system provides an example of organ-specific growth regulation, in that gibberellin, the key to the food reserves, is made in the embryo, which also contains the only tissues in the seed capable of growth.

How does gibberellin cause the appearance of α-amylase activity? First, it is clear that the enzyme is formed de novo from its constituent amino acids and is not simply an activated form of a previously synthesized

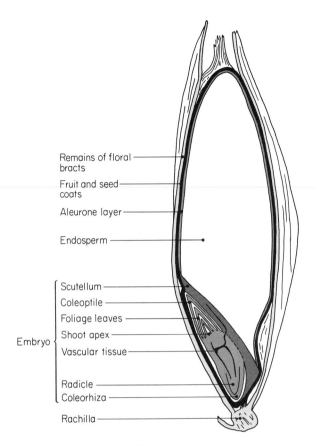

Remains of floral bracts

Fruit and seed coats

Aleurone layer

Endosperm

Embryo {
Scutellum
Coleoptile
Foliage leaves
Shoot apex
Vascular tissue

Radicle
Coleorhiza
}

Rachilla

Fig. 4-12   Longitudinal section of a barley grain.

inactive stored protein. This was shown by the fact that the addition of radioactive amino acids to the barley grains or aleurone layers incubated with gibberellin resulted in the incorporation of radioactivity into the protein. This incorporation was prevented by inhibitors of protein synthesis such as cycloheximide. The point in the protein synthesis at which gibberellin acts is indicated by the fact that inhibitors of DNA-dependent RNA synthesis (e.g., actinomycin D) also prevent amylase synthesis (3) (Fig. 4-15). Therefore, it is implied that gibberellin must be involved in the production of messenger RNA molecules on the DNA template. Gibberellin is probably acting as a derepressor of those genes that code for the hydrolytic enzymes, thus allowing the production of these enzymes, which were previously "turned off" by repressors within the nucleus. We can gain some idea of the time course of the production of amylase resulting from gibberellin action by the time taken to shut down the

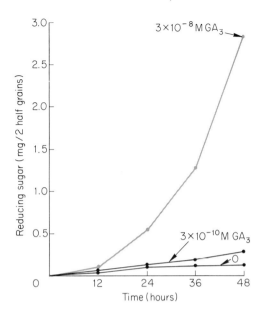

Fig. 4-13 The production of reducing sugars during the incubation of embryonectomized barley grains at 20°C is greatly increased in the presence of gibberellin through the induction of α-amylase. [From data of Coombe, et al. (1).]

Fig. 4-14 The enhancement of α-amylase production in embryoless barley grains, as seen by the increase in the formation of sugars from starch, occurs only in the presence of the aleurone cells, which are the site of enzyme synthesis. (From data of MacLeod and Millar in van Overbeek. 1966. *Science* **152**:721–731.)

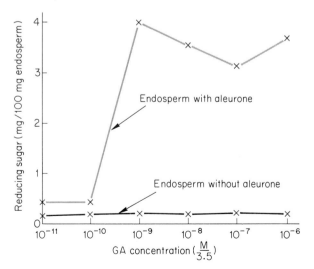

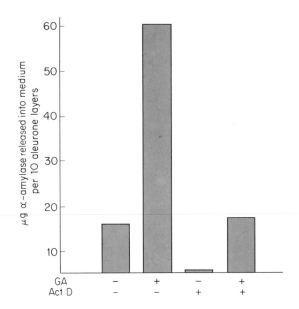

Fig. 4-15 The enhanced production of α-amylase by isolated aleurone layers in the presence of gibberellic acid is prevented in the presence of actinomycin D. [From data of Varner and Chandra (3).]

Fig. 4-16 The addition of inhibitors to isolated aleurone layers in the presence of gibberellic acid (0.05/0.1 μM) curtails the further production of α-amylase. This effect is manifested more rapidly in the presence of a protein synthesis inhibitor (cycloheximide) than in the presence of an RNA synthesis inhibitor (6-methylpurine), giving some measure of the longevity of the gibberellin-promoted m-RNA coding for α-amylase. [From data of Chrispeels and Varner (4).]

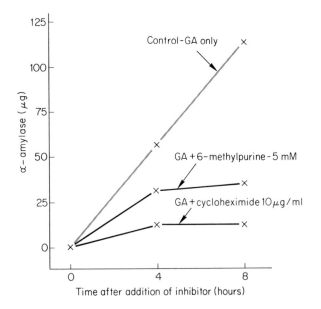

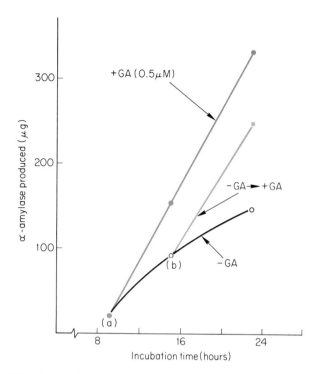

Fig. 4-17   Gibberellin must be continuously present for the production of α-amylase by isolated aleurone layers. The removal of GA at (a) caused a slow decrease in the rate of amylase production, but if GA is replaced at (b), the original rate of amylase production is immediately restored. [From Chrispeels and Varner (4).]

synthesis when inhibitors are added after the GA. Protein synthesis inhibitors had an almost immediate effect, as would be expected, while the effect of RNA synthesis inhibitors was delayed for 3 to 4 hours, indicating that this is the approximate life of the synthesized amylase m-RNA (Fig. 4-16) (4).

In the barley aleurone system the gibberellin has to be continually present to maintain the formation of amylase. Removal of the hormone in the mid-course of amylase production results in a slowing down of amylase synthesis, but if the gibberellin is replaced, then the original rate of enzyme production is reestablished without any lag period (Fig. 4-17). This is just the way an on-off repressor system is envisaged to act, though the data can also be interpreted as control at the transcription level, with a requirement for m-RNA production for the hormone to act.

In addition to α-amylase, gibberellin has been shown to initiate the formation of other hydrolases, notably protease and ribonuclease, so the

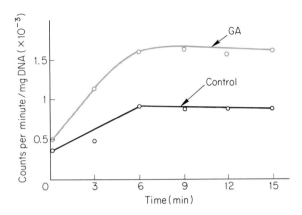

Fig. 4-18   The presence of gibberellin enhances the incorporation of cytidine triphosphate-³H into RNA in isolated nuclei. [From Johri and Varner (5).]

one hormone appears to render all the food reserves of the seed available to the young plant.

Further information relative to the above hypothesis has been obtained by examining the effect of GA on RNA synthesis in isolated nuclei (5). Nuclei are difficult to isolate from aleurone cells, so pea stem nuclei were used instead. Here gibberellin was found to increase the RNA formation by these nuclei, particularly m-RNA (Fig. 4-18). The RNA synthesized by the GA-treated nuclei was also qualitatively different from the non-hormone-treated nuclei, indicating that gibberellin had indeed altered the genes that were producing RNA. Perhaps in the not-too-distant future we shall know the mechanism by which gibberellin changes the RNA production and also the exact nature of the RNA's whose production was induced by gibberellin.

This explanation of gibberellin action provides the most complete story for plant hormone action at the present time, but it is clearly not the entire answer as yet. For example, the most conspicuous effect of gibberellin is in increasing stem growth. Does this come about in the same manner? Little work has been done on this particular problem but it seems probable that the mechanism described for the aleurone applies in other tissues as well. Let us recall, for example, that the isolated nuclei used on the above studies came from pea stems, and other experiments show that nucleic acid and protein synthesis are also involved in gibberellin-induced stem growth. The stem enzymes ultimately controlled by gibberellin are probably different from the aleurone example. One possibility is peroxidase, whose activity is very high in many dwarfs and dramatically lowered by GA_3 application (6).

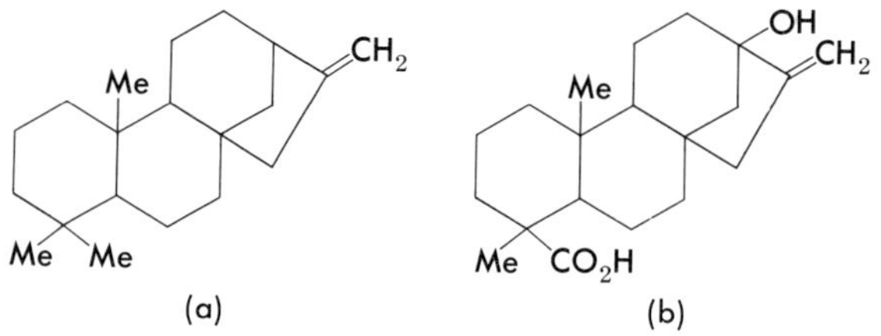

Fig. 4-19   The diterpenes (-)-kaurene (a) and steviol (b) are possible precursors of gibberellin in higher plants. Kaurene is known to be the precursor of gibberellic acid in *Gibberella fujikuroi*.

The biosynthesis of gibberellin

Gibberellin is an isoprenoid compound formed from diterpenes, which are well-known angiosperm metabolites. In higher plants the immediate precursors of gibberellin are believed to be ($-$) kaurene or steviol, probably varying according to the species (7, 8) (Fig. 4-19). When steviol is supplied to plants, it is converted to gibberellin. The pathway of the diterpenes, kaurene and steviol, in gibberellin biosynthesis, which was elucidated by feeding ^{14}C-labelled compounds to the plants appears to be

$$\text{Acetate} \rightarrow \text{mevalonate} \rightarrow \text{diterpene} \rightarrow \text{gibberellin}$$

The identification of diterpenes as intermediates in gibberellin biosynthesis is further supported by the fact that AMO-1618, a growth retardant that acts as a gibberellin antagonist, prevents the cyclization of the diterpene precursor (geranylgeranylpyrophosphate) into either kaurene or steviol. Compounds such as helminthosporol, which lack large portions of the gibberellin structure but are capable of mimicking gibberellin activity, are also isoprenoid in origin.

Summary

Gibberellins, originally discovered as products of a pathogenic fungus, are now known to be natural plant hormones. Over 20 different gibberellins exist. They enhance stem and leaf growth, promote bolting and flowering in some long-day plants, induce parthenocarpic fruit set, and are involved in the breaking of dormancy. The difference between genetically dwarf and tall plants may involve deficient gibberellin production in the dwarf plants. Many of the responses to gibberellin application are in systems normally controlled by phytochrome or induced by chilling. These treatments may act through induction of gibberellin synthesis. The actions of gibberellin and auxin are frequently synergistic.

In cereal grains, gibberellin induces the production of α-amylase, which hydrolyzes the reserve starch in the endosperm. Both this enzyme production and promotion of stem or leaf growth can be used for a gibberellin assay. In amylase production gibberellin probably acts by derepression of the genes coding for the enzyme. Gibberellin biosynthesis from the isoprenoid precursor mevalonate occurs via the ditepenoid compound steviol.

REFERENCES

General

Brian, P. W. 1966. The gibberellins as hormones. Int. Rev. Cytol. **19:**229–266.

Paleg, L. G. 1965. Physiological effects of gibberellins. Ann. Rev. Plant Physiol. **16:**291–322.

van Overbeek, J. 1966. Plant hormones and regulators. Science **152:**721–731.

1. Coombe, G. B., D. Cohen and L. G. Paleg. 1967. Barley endosperm bioassay for gibberellins. I. Parameter of the response system. Plant Physiol. **42:**105–112.

2. Ockerse, R. and A. W. Galston. 1967. Gibberellin-auxin interaction in pea stem elongation. Plant Physiol. **42:**47–54.

3. Varner, J. E. and G. R. Chandra. 1964. Hormonal control of enzyme synthesis in barley endosperm. Proc. Nat. Acad. Sci. U.S. **52:**100–106.

4. Chrispeels, M. J. and J. E. Varner. 1967. Hormonal control of enzyme synthesis: On the mode of action of gibberellic acid and abscisin in aleurone layers of barley. Plant Physiol. **42:**1008–1016.

5. Johri, M. M. and J. E. Varner. 1968. Enhancement of RNA synthesis in isolated pea nuclei by gibberellic acid. Proc. Nat. Acad. Sci. U.S. **59:**269–276.

6. McCune, D. C. and A. W. Galston. 1959. Inverse effects of gibberellin on peroxidase activity and growth in dwarf strains of peas and corn. Pl. Physiol. **34:**416–418.

7. West, C. A., M. Oster, D. Robinson, F. Lew and P. Murphy. 1969. Biosynthesis of gibberellin precursors and related diterpenes, p. 313–332. *In* F. Wightman and G. Setterfield [eds.] Biochemistry and physiology of plant growth substances. Runge Press, Ottawa.

8. Ruddat, M. 1969. Biosynthesis and metabolism of steviol, p. 341–346. *ibid.*

FIVE

Cytokinins

One of the most intriguing problems in biology involves the regulation of the cell division cycle. Why do cells divide at one time and stop at another time? What causes meristematic cells to retain their ability to divide? And what, in turn, regulates the failure of other cells ever to divide in the mature organism? We know that in animals aberrations in the control of cell division lead to the formation of tumors, i.e., masses of cells produced in opposition to the normal controls regulating form. Some such tumors can be malignant and can induce yet other cells to resume cell division activity long after it has normally ceased. Such malignancies, characteristic of the disease called cancer, may so upset regulatory mechanisms in the organism that death results. Clearly, not only integration of the organism but survival itself demands that there be careful regulation of the cell division machinery.

A particularly striking example of the regulation of cell division activity and of the complexities of the regulatory system may be seen from experiments with animals in which portions of mature organs are extirpated and regeneration is allowed to occur. We know, for example, that in rats the liver reaches a certain stable size that is characteristic of the organism and is related to the size of other organs in the body. If one removes surgically a portion of a mature liver that has ceased all cell division, the cells on the cut surfaces resume cell division activity. New cells are produced in such quantity as to replace those that have been experimentally

115

removed and these new cells then differentiate into normal functioning liver cells, thus repairing the structural damage. Obviously, at least two independent control mechanisms are at work here; one control, restraining cell division, shuts off the mitotic activity of the cells that originally gave rise to the liver, and the second, activated by injury and depletion of the organ, causes resumption of the previously arrested mitotic activity. The normal shutoff mechanism then presumably once again intervenes to bring the new cell division activity to a halt.

Many similar phenomena can be seen in plants. For example, the cells of a potato tuber do not usually divide. Yet if a tuber is cut open and permitted to remain in a moist atmosphere, the cells just below the cut surface resume mitotic activity and produce several layers of a closely packed, ultimately corky tissue that effectively seals the cut surface. The same phenomenon may be seen in the external layers of the cortex of a young tree when the secondary growth, produced by a cambium, causes splitting of the outer surface. New cork cambia arise below the split and give rise to periderm tissues similar to that formed in the cut potato tuber. Here again we have an interplay of two regulatory mechanisms, one normally leading to cessation of cell division and the other, initiated by injury, which permits the resumption of mitotic activity.

Plants also may form tumors of either the benign or the malignant type. Benign (limited) tumors, frequently noted on the trunks of trees, may be formed as a result of the localized promotion of cambial activity by stress or injury. Malignant tumors frequently result from the invasion of the plant system by microorganisms. One example of this is the crown-gall disease, initiated when the bacterium, *Agrobacterium tumefaciens*, enters plants through an injury. A primary tumor, formed at the site of the inoculation, contains recoverable viable bacteria. The tumor is composed of cells proliferating in all directions in an uncontrolled fashion, producing a knobby or knotty swelling (Fig. 5-1). Subsequently, at other sites on the plant, secondary tumors arise that are frequently bacterium-free. This implies that the bacterium is not required for continuation of the tumorous condition; rather, it serves merely as a convenient package for introducing some other entity, possibly a virus, into the plant. Primary or secondary tumors, or portions of stem containing them, can transmit the disease when grafted onto receptors. Thus it appears that a "tumor-inducing principle" initially elaborated by crown-gall bacteria in susceptible plant cells can induce an atypical unregulated growth of such cells. Although the mechanism by which the crown-gall bacterium acts is not at all well understood, the properties of the altered cells have been extensively examined in tissue culture. Whereas normal cells require the addition of exogenous auxins and other growth factors, crown-gall cells frequently need no such addenda, and in fact are some times even inhibited when auxin or other

Fig. 5-1 A crown gall tumor produced on a geranium stem by inoculation with *Agrobacterium tumefaciens*. (Courtesy of G. Morel and R. J. Gautheret, France.)

growth-regulating materials are added from the outside. The tumefaction process seems to have "turned on" numerous latent synthetic abilities of the infected cells.

Certain other diseases and transformations of plants, ranging from the formation of nodules on leguminous roots (Fig. 5-2) to fasciations and witches' brooms (Fig. 5-3), can be shown to be correlated with the inception of cell division activity in previously nondividing cells (1). In many instances there is physiological evidence for the existence of chemical substances responsible for this activity (2). There is thus adequate background from various sources for believing in substances that have the capacity to regulate the division of plant cells.

The experiments that led to the final isolation and identification of plant cell-division-promoting substances (*cytokinins*) stemmed from attempts to induce the development in vitro of two kinds of plant cells: (a) the unfertilized ovules of *Datura stramonium*, the Jimson weed, upon which much genetic research has been carried out, and (b) the cells of the pith of the tobacco plant, which never divide *in situ*. We shall describe each of these cases in some detail.

Fig. 5-2   Nodules on soybean roots. (Courtesy of The Nitragin Company.)

Datura eggs and coconut milk

Datura stramonium was, for a long time, a favorite plant for investigation by geneticists and cytologists interested in the effects of chromosomal inversions, deletions, reciprocal translocations, extra chromosomes and portions thereof. For much of the genetic work necessary for analysis of these problems, homozygous, pure strains were desirable. Breeding to produce such homozygous plants was a long and tedious process, and it occurred to van Overbeek and coworkers (3) to attempt to stimulate directly the development of the unfertilized egg from the embryo sac. If such a cell could be induced to divide, it might produce either a viable haploid plant or a diploid formed by an abnormal chromosome-doubling process. In either event, it would provide a source of homozygous plant material. The technique involved in these experiments was to excise aseptically the contents of the embryo sac and to transfer them to nutrient media designed to make them grow. It was immediately apparent that although young embryos of the "torpedo" stage and beyond could be made to develop, unfertilized eggs and very small embryos could not. In fact, there seemed to be a regular progression from a completely heterotrophic situation in the unfertilized egg to greater autotrophism in successively older embryos. In an attempt to find a natural source that might contain all of the factors required for the growth of the unfertilized egg, these researchers made use of the liquid endosperm of the coconut, generally referred to as coconut milk. This material, which contains many free-floating nuclei in a remarkably rich medium, becomes solid later on as each nucleus becomes a center of activity leading to the formation of cell walls around it. This process forms the solid white endosperm well known as a food. Coconut milk, although still not able to make the unfertilized egg divide, could promote the growth of embryos so young that no combination of known growth factors could stimulate their growth. Clearly, then, coconut milk must contain some substance

Fig. 5-3 A "witches'-broom" on a willow tree resulting from the growth of many lateral buds (compare regular branches upper right) caused by cytokinin production by an invading plant pathogen. (Courtesy of W. A. Sinclair, Cornell University.)

Fig. 5-4   The vast array of slowly rotating bottles used by F. C. Steward and colleagues in investigating the effects of plant growth substances on carrot cells cultured in vitro. (Courtesy of F. C. Steward, Cornell University.)

or substances responsible for this unique action. This realization led several researchers to initiate attempts to isolate and characterize the hypothetical cell division factors.

F. C. Steward and colleagues at Cornell University employed strains of carrot in their attempts to detect and assay such substances (Fig. 5-4). Carrot had been shown many years earlier to initiate callus growth in the presence of auxin, and it soon became clear that coconut milk could greatly enhance the rate at which cell divisions occur in such tissues (4).

After many years of investigating coconut milk and the liquid endosperms of other seeds, such as *Aesculus,* Steward and colleagues isolated *myo*-inositol, several auxin conjugates and various amino acid fractions which they could show contributed to the growth-promoting properties of the liquid endosperm. However, no single material was found that could duplicate the major action of coconut milk.

Tobacco pith and the discovery of kinetin

Another approach was adopted by Skoog and collaborators at the University of Wisconsin. They excised aseptically portions of the nondividing mature pith from the interior of stems of well-developed tobacco plants. Such explants had a limited ability to enlarge by cell expansion under the influence of auxin and the other components of standard nutrient media. In surveying known materials for their ability to induce cell division in this system, Skoog found that the purine adenine [Fig. 5-5(a)] and its nucleoside, adenosine, could lead not only to cell division and more prolonged growth, but to the promotion of bud development as well (5).

In subsequent experiments, nucleic acids in which adenine occurs as a component were examined. One sample of herring sperm DNA was found to yield results quantitatively much superior to those produced by adenine. When the phenomenon was examined further, it was found that not all samples of the DNA produced this result. Later, it was found that any sample of DNA, or, in fact, even adenosine could be converted to an extremely active material merely by autoclaving it under appropriate conditions. The new substance formed under these stringent conditions was isolated and identified as 6-furfuryladenine (6). Because of its great activity in inducing cell division when applied together with auxin, it was named *kinetin* [Fig. 5-5(b)]. The availability of this synthetic material naturally opened up great new fields for investigation.

Actions of kinetin

Synthetic kinetin, which, it should be emphasized, does not occur naturally in the plant, was found to interact with auxin in a wide variety of physiological processes. When added together with auxin, kinetin can initiate cell divisions and can affect such morphogenetic phenomena as bud initiation in excised tobacco callus in a similar fashion to adenine. The resultant differentiation in tissue culture depends on the relative

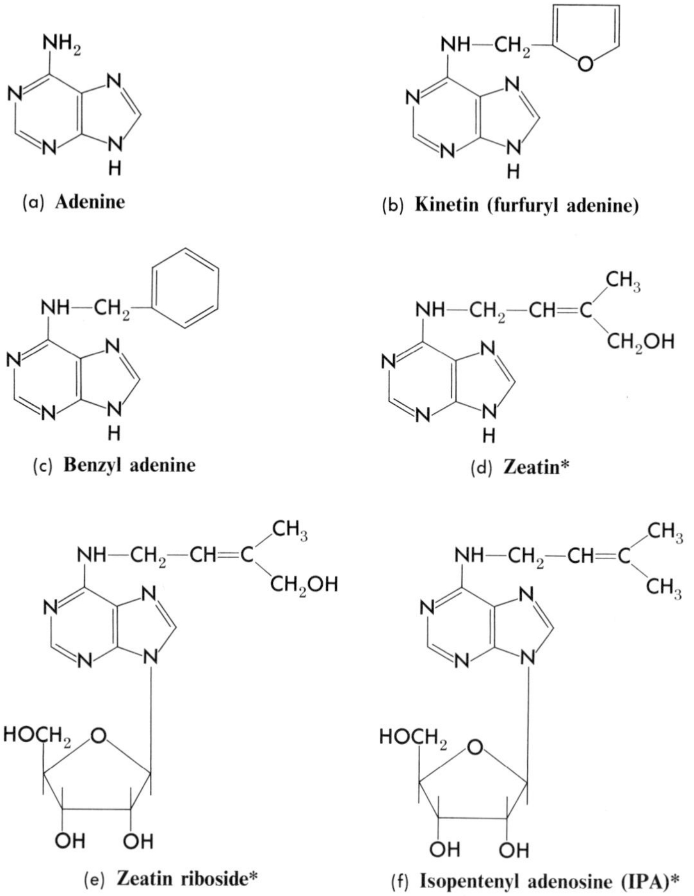

(a) Adenine

(b) Kinetin (furfuryl adenine)

(c) Benzyl adenine

(d) Zeatin*

(e) Zeatin riboside*

(f) Isopentenyl adenosine (IPA)*

Fig. 5-5   The structural formulae of various adenine derivatives with cytokinin activity. The starred compounds are natural cytokinins.

concentrations of the growth substances present. High molar ratios of kinetin to auxin were found to lead to the formation of buds; at roughly equal concentrations of auxin and kinetin, callus growth was produced exclusively, and when the ratio of auxin to kinetin was high, there was a tendency for root growth to be initiated (Fig. 5-6) (6). In typical cell elongation systems such as oat coleoptile and pea epicotyl extension growth, kinetin tends to inhibit auxin-stimulated longitudinal growth and to promote transverse growth. The interaction with auxin necessitated a systematic investigation of other physiological systems, such as apical dominance, known to be influenced by auxin. Here it was found that the local application of kinetin to repressed buds caused a release from inhibi-

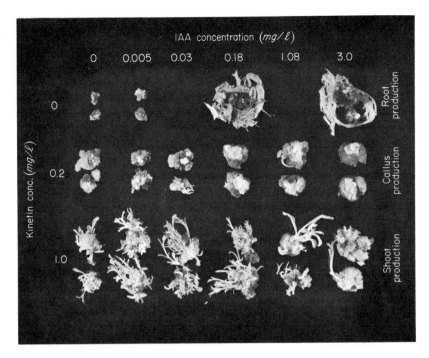

Fig. 5-6   The interactions of kinetin and IAA on callus production and tissue differentiation in tobacco tissue culture. Maximum tissue production requires the presence of auxin and kinetin. High ratios of auxin/kinetin produce roots; high ratios of kinetin/auxin produce shoots; at equal concentrations callus predominates. [From Skoog and Miller (6).]

tion (Fig. 5-7) (7). In fact, quantitative experiments revealed that one could essentially titrate kinetin against auxin. The more auxin coming from the apex, the more inhibited was the bud, but the more kinetin supplied to the bud, the more auxin was required to inhibit it. This auxin-cytokinin antagonism could also be used to explain the formation of "witches' brooms," which result essentially from a complete removal of the inhibition normally placed on dormant buds separated from the apex only by a few short internodes. Presumably this happens naturally when the fungi that induce witches' brooms invade the plant and produce substances with kinetin activity.

Because kinetin does not occur naturally, diverse chemical substances with kinetinlike activity were at first collectively called kinins. Later, when it was recognized that this name had previously been employed by animal physiologists to describe polypeptides that control contraction of smooth muscle, the name was changed to *cytokinins.*

Fig. 5-7   The release of lateral buds from apical dominance by kinetin. The plant on the right was treated with a local application to the bud of 330 ppm kinetin 3 days previously. [From Sachs and Thimann (7).]

Cytokinins and leaf senescence

Cytokinins turned out to explain another phenomenon that had long puzzled plant physiologists. It had frequently been observed that when leaves, such as those of tobacco, are removed from the plant, their protein content diminishes rapidly, while soluble nitrogen rises. This massive breakdown of protein was thought to account, at least in part, for the short life span of many cut plants and plant parts, especially leaves. It was fortuitously discovered that the incorporation of kinetin into the nutrient solution bathing the petiole of excised *Xanthium* leaves caused a prolonged retention of the green color of the leaf (8). Thus, kinetin seemed to have an antisenescence effect, which was later shown to be due to its maintenance of protein and nucleic acid synthesis, and therefore its regulation of the protein-soluble nitrogen equilibrium (9) (Fig. 5-8).

In this action, the influence of kinetin was shown to be restricted to the locale to which it was applied. So, for example, if kinetin was applied to one-half of a leaf, only that half remained green after prolonged excision. If kinetin was applied only to a small circle of tissue, that circle remained green, while the rest faded. It could also be shown that these persistent islands of kinetin-induced greenness became centers for the accumulation of labeled sugars, amino acids, and inorganic ions applied at distant loci (10) (Figs. 5-9 and 5-10). Though these phenomena could largely be correlated with kinetin-induced protein synthesis at the locale of application, this is not the entire story, as aminoisobutyric acid, an amino acid that is not incorporated into protein, also moves into cytokinin-treated areas.

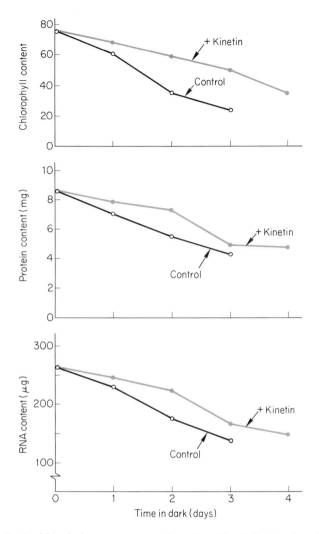

Fig. 5-8 Kinetin delays leaf senescence, as evidenced by chlorophyll destruction, due to the maintenance of nucleic acid and protein synthesis. In this experiment excised leaf discs of Xanthium were floated on a solution of water or 40 mg/1 kinetin in darkness, and the changes during senescence followed. [From Osborne (9).]

Naturally occurring cytokinins

So many interesting physiological activities of kinetin were being discovered that it became necessary to examine plant materials for naturally occurring cytokinins. Several synthetic analogs of kinetin had already been found to be active. One such, benzyladenine [Fig. 5-5 (c)], had even

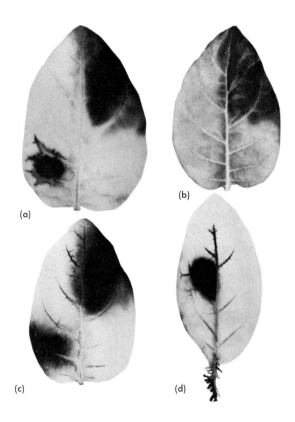

Fig. 5-9 The effect of kinetin on the accumulation of locally applied radioactively-labeled materials in excised tobacco leaves. (a) Radioautogram. Kinetin applied upper right, labelled glycine, lower left. Dark areas shows where label exists. (b, c) Kinetin applied upper right, labelled α-amino-butyric acid, lower left. (b) is a photograph showing persistent green color in kinetin-treated area; (c) is a radioautogram of the same leaf showing accumulation of label. (d) radioautogram of a rooted leaf of tobacco to which labelled α-amino butyric acid has been applied to the middle left. Note the accumulation of label in the roots and along the veins. This implies a synthesis of cytokinin in the roots and its movement upward through the vascular system. (Courtesy of K. Mothes, Halle, Germany.)

found use commercially in prolonging the usable life of certain harvested green crops, such as broccoli heads. The workers who began the investigations on natural cytokinins made the reasonable assumption that the naturally occurring materials would at least to some extent resemble kinetin in their chemistry. This guided the formulation of suitable isolation schemes. The best sources turned out to be young fruits of apples and pears, in which cell division was proceeding vigorously, and germinating seedlings, in which abundant cell division activity was occurring in many regions. From many such materials fractions could be obtained which, when applied together with auxin, stimulated cell division in suitably

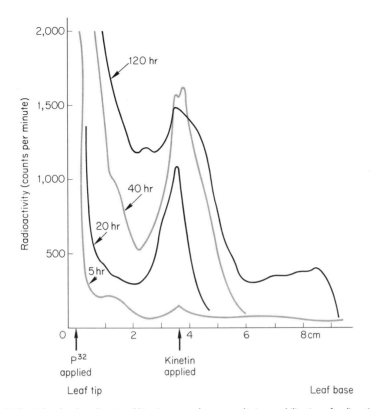

Fig. 5-10 A localized application of kinetin to corn leaves results in a mobilization of radioactivity from $^{32}PO_4$ applied near the leaf tip. Scanning the leaf at intervals reveals an accumulation of radioactivity at the position of kinetin application. (From Leopold. 1964. Plant growth and development. McGraw-Hill, New York. After Mueller, 1964).

explanted tobacco pith cells. Soon it was possible to report that naturally occurring substances with cytokinin activity did exist and in 1964 Letham and his associates in New Zealand announced the isolation and structure of a cytokinin that occurs in young maize seeds. This material, named *zeatin* [Fig. 5-5 (d)], is, like kinetin, an adenine derivative, bearing a 5-carbon isoprenoid substituent instead of the furfuryl group on the nitrogen of the 6-amino group. Therefore, it appears that, as with gibberellins, there exists a family of active molecules differing slightly in structure. The latest member of the family is zeatin-riboside [Fig. 5-5 (e)], a ribose sugar derivative of zeatin, which has been found to occur in sweet corn. At long last, in 1967, 26 years after the growth-promoting properties of coconut milk were discovered, its major active constituent was shown also to be zeatin-riboside (11).

In plants, the functions of the naturally occurring cytokinins are manifold. They appear to be synthesized in the root apex and are translocated via the xylem to the shoot. This has been demonstrated by the fact that cytokinin activity can be found in the bleeding sap of various species, particularly grapes, when the stem is removed just above the roots. Once in the leaves the cytokinins play an important role in the regulation of metabolism and senescence. In young fruits and developing seeds they also clearly stimulate cell division and growth.

With the elucidation of the family of cytokinins as adenine derivatives, and the discovery of kinetin originating from autoclaved DNA, it was logical to question whether cytokinin molecules might occur as integral parts of nucleic acids. This field has now been widely surveyed, and cytokinin activity has in fact been detected in hydrolyzates of RNA from many species, including plants, microorganisms (yeast and *E. coli*) and animal cells. The occurrence of the cytokinins in RNA was, however, very specific, being limited to transfer RNA hydrolyzates and never in hydrolyzates of ribosomal RNA. The first naturally occurring identified cytokinin to be found as an integral part of a nucleic acid was N^6 (Δ^2-isopentenyl) adenosine (IPA) [Fig. 5-5 (f)] from yeast t-RNA. The same material was later identified in spinach and peas, while the hydroxylated derivative (zeatin-riboside) was isolated from the t-RNA of immature sweet corn kernels (12).

On further examination it was found that the occurrence of cytokinins was even more specific, being limited to certain t-RNA's. It was calculated that one active molecule per 20 t-RNA's would be sufficient to account for the measured activity in the total t-RNA. Activity was found in hydrolyzates of serine, isoleucine and tyrosine t-RNA, but none in arginine, glycine, phenylalanine, or valine t-RNA. Analytical studies of the nucleotide sequence of serine t-RNA showed that IPA was an integral part of the molecule, occurring in only one specific position adjacent to the triplet of bases that formed the anticodon of the t-RNA (Fig. 5-11). The lack of cytokinin activity in some transfer RNA hydrolyzates was also explained by the absence of IPA in alanine and phenylalanine t-RNA nucleotide sequences.

The mode of action of cytokinins

Clearly, if one is looking for a single molecular event that underlies all the varied activities of the cytokinins, one should be guided by their demonstrated promotion of protein synthesis, which is in turn known to be controlled by nucleic acids. The occurrence of cytokinin molecules in t-RNA looked like an obvious key to the way in which cytokinins might exert their effects on protein synthesis. The answer seemed to be clear when

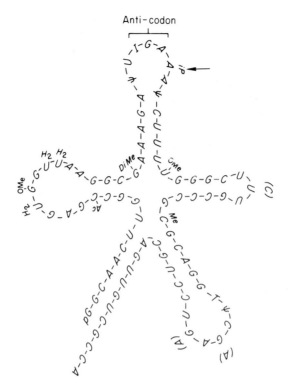

Fig. 5-11 The structure of serine transfer RNA showing the position of IPA (arrowed) adjacent to the anticodon (bracketed).

it was found that the IPA adjacent to the anticodon possessed an important function, namely, the attachment of the t-RNA to the ribosome-messenger complex. If the structural integrity of the IPA was altered, binding was prevented. The presence or absence of cytokinin would, therefore, be important for the promotion and functioning of several t-RNA's, and through this means it might control protein synthesis. Though control of the type of protein produced is generally thought to be invested in the type of messenger RNA being produced at any given stage in development, evidence has accumulated that in addition, t-RNA's may exert some control in the overall system. Although this has been known in bacteria for some time, information about higher plants is scanty. Some changes in t-RNA populations during differentiation in wheat seedlings have been described (13) and, though this evidence is not conclusive, it does indicate that t-RNA control of development in plants is at least possible.

Further evidence that this might be the role of cytokinin was obtained when it was found that ^{14}C-labeled benzylaminopurine supplied to soybean and tobacco tissue cultures, which require cytokinin for growth, could be recovered in the t-RNA extracted from the tissues, mainly as the nucleotide of 6-benzylaminopurine. Fractionation of the t-RNA revealed that the labeled material was found in only one subfraction of t-RNA, suggesting that cytokinins were preferentially incorporated into certain t-RNA's (14), in agreement with the now-known t-RNA structures. It would thus seem that the case for the action of cytokinins was closed; cytokinins could be envisaged as being incorporated into certain t-RNA's and thereby influencing changes in protein synthesis and development.

The above hypothesis has, however, recently been questioned on several grounds. First, chemists have shown that the various "minor bases" in t-RNA, such as the methylated purines and IPA, are not incorporated into the molecule as such. Rather, a regular nucleotide is structurally rearranged and methyl or other side chains are added on after the polymeric molecule has been completed. If this is generally true, then cytokinin molecules should not be incorporated into the t-RNA *per se*, but should be formed within the t-RNA by addition of the isopentenyl side chains. This indicates that exogenously applied cytokinin could not function by being directly incorporated into t-RNA. In support of this notion, recent results of Hall and co-workers (15) indicate that the IPA in t-RNA results from the attachment of an isopentenyl group derived from mevalonic acid to a specific adenosine residue of the preformed t-RNA, and enzyme systems that perform this attachment have been isolated. The presence of IPA in the media containing mevalonic acid did not have any competitive effects on the incorporation of mevalonic acid (15), while methylation at the 9-position of the purine ring of 6-benzylaminopurine fed to tissue culture prevented its incorporation into t-RNA and yet did not alter its cytokinin activity (16). This suggested that the previously reported incorporation of benzyladenine into t-RNA was in fact incorporation of the adenine nucleus or the side chain separately following degradation of the benzyladenine. It is also interesting to note that ethanolic extracts of corn kernels contain the *trans*-isomer of zeatin while the t-RNA hydrolyzates of the same source yield the *cis*-isomer, further suggesting that zeatin is not a precursor in the synthesis of t-RNA.

If cytokinins are incorporated per se into t-RNA, then as cytokinin activity has been shown to be present in *E. coli* t-RNA, it would be anticipated that a cytokinin-requiring mutant of *E. coli* could be found. This would be a bacterium that through mutation had lost the ability to synthesize the required cytokinin molecule from its precursors. Although over 10,000 colonies were examined (16) no cytokinin-requiring mutants were found, though many appeared which required other nutrient supplements. It was calculated that the possibility that a cytokinin auxotroph existed was about one in 10^{-105}, i.e., very small indeed. This indicated that cytokinin was probably not a requirement for bacterial growth and by inference not required for the incorporation into t-RNA in the organism.

Though incorporation of applied cytokinins per se into t-RNA now seems unlikely as their mode of action, can this still be the way in which cytokinins function following degradation of applied cytokinin? If this were so, then mevalonic acid should be able to substitute for cytokinins. In tobacco pith tissue mevalonic acid can indeed partially replace the requirement for a cytokinin (15). There is still a further requirement for cytokinin, however, so though it is possible that part of the action of applied cytokinin concerns breakdown followed by incorporation into t-RNA, cytokinins must still possess alternative modes of action.

It has been suggested that the naturally occurring free cytokinins in plants are breakdown products, rather than precursors, of t-RNA molecules. This cannot be the entire story, as the content of cytokinins in some tissues far exceeds that which could be derived from t-RNA breakdown. In addition, this would not explain the need for exogenously supplied cytokinin in cultured cells unless, as seems unlikely, a block to t-RNA breakdown exists in these cells. Cytokinins are therefore probably made directly from mevalonate as well as being formed by t-RNA breakdown.

Enzyme systems have been isolated which catalyze the degradation of IPA. Although the functions of such a degradative system are obscure, they may be of importance to cytokinin action as, by comparison, many other minor components of t-RNA catabolism are secreted from the cells unchanged. In mammalian systems the degradative enzymes are highest in rapidly dividing and differentiating cells, but it is not known whether this is true in plants.

One further piece of evidence argues against the action of cytokinins as being due to their presence in t-RNA. When moss protonemata are treated with cytokinins, they are induced to form buds and labeled cytokinins can be located only in those cells differentiating into buds. When the protonemata are, however, removed to a non-cytokinin-containing medium, the labeled cytokinin is washed out of the "target" cells and the developing bud initials revert to their former state (Fig. 5-12) (17). It thus seems that free cytokinin is not acting as a trigger mechanism but must be present throughout the development of the buds. This fact suggests a loose, reversible binding to a receptor site rather than incorporation into t-RNA as a mode of action. The nature of the binding or the receptor is, however, unknown at present. It is possible that the mechanism of bud formation on protonemata is different from that governing the division of cells in tissue culture. Nevertheless it is clear that direct incor-

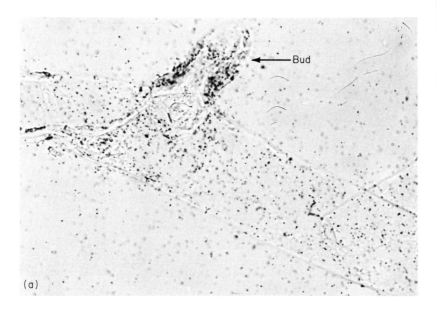

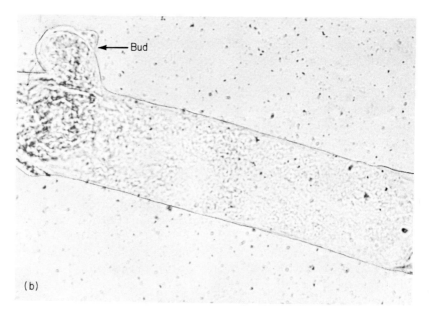

Fig. 5-12 Moss (*Funaria*) filaments treated with benzyladenine-^{14}C show an accumulation of radioactivity (black dots) in the buds after a 10 to 12 hour treatment with benzyladenine. (a) If, however, this treatment is followed by a 3-minute rinse in water, the benzyladenine is completely removed (b), showing that the benzyladenine is not permanently incorporated into some substance in the buds. [From Brandes and Kende (17).]

poration of cytokinin molecules into t-RNA cannot be the complete answer to the way in which cytokinins act and that we must continue to search for the exact mode of action of these interesting substances.

Summary

Cell division factors were originally found in coconut milk. In tobacco pith culture adenine derivatives were found to be active, the most potent being 6-furfuryladenine or *kinetin*. This compound, which does not occur naturally, caused cell division in tissue culture, release of buds from apical dominance, delay of leaf senescence by maintenance of protein and nucleic acid synthesis, and creation of a sink for metabolites. *Cytokinins* have been found to occur naturally in plants, particularly in young fruits and germinating seedlings. The first such compound isolated was termed *zeatin*, and the active constituent of coconut milk has been identified as *zeatin riboside*. Isopentenyl adenosine also occurs naturally.

Cytokinin molecules have been found to occur in t-RNA adjacent to the anticodon, where they are involved in the attachment of the t-RNA to the ribosome-mRNA complex during protein synthesis. This may not be the mode of action of applied cytokinins, however, as the formation of cytokinin in the t-RNA occurs by modification of the adenine already in the t-RNA and applied cytokinins are not incorporated into t-RNA. The mode of action of applied cytokinins is unknown at the present time.

REFERENCES

General

Helgeson, J P. 1968. The cytokinins. Science **161**:974–981.

Galston, A. W. and P. J. Davies. 1969. Hormonal regulation in higher plants. Science **163**:1288–1297.

Letham, D. S. 1967. Chemistry and physiology of kinetin like compounds. Ann. Rev. Plant Physiol. **18**:349–364.

Letham, D. S. 1969. Cytokinins and their relation to other phytohormones. BioScience **19**:309–316.

Miller, C. O. 1961. Kinetin and related compounds in plant growth. Ann. Rev. Plant Physiol. **12**:395–408.

1. Braun, A. C. and T. Stonier. 1958. Morphology and physiology of plant tumors. Protoplasmatologia **10**:5a, 93 p.

2. Thimann, K. V. and T. Sachs. 1966. The role of cytokinins in the "fasciation" disease caused by *Corynebacterium fascians*. Amer. J. Bot. **53**:731–739.

3. van Overbeek, J., M. E. Conklin and A. F. Blakeslee. 1941. Factors in coconut milk essential for growth and development of very young *Datura* embryos. Science **94**:350–351.

4. Steward, F. C. and E. M. Shantz. 1959. The chemical regulation of growth (some substances and extracts which induce growth and morphogenesis). Ann. Rev. Plant Physiol. **10**:379–404.

5. Skoog, F. and C. Tsui. 1948. Chemical control of growth and bud formation in tobacco stem segments and callus cultured in vitro. Amer. J. Bot. **85**:782–787.

6. Skoog, F. and C. O. Miller. 1957. Chemical regulation of growth and organ formation in plant tissues cultured in vitro. Symp. Soc. Exp. Biol. **11**:118–131.

7. Sachs, T. and K. V. Thimann. 1967. The role of auxins and cytokinins in the release of buds from dominance. Amer. J. Bot. **54**:136–144.

8. Richmond, A. E. and A. Lang. 1957. Effect of kinetin on protein content and survival of detached *Xanthium* leaves. Science **125**:650–651.

9. Osborne, D. J. 1962. Effect of kinetin on protein and nucleic acid metabolism in Xanthium leaves during senescence. Plant Physiol. **37**:595–602.

10. Mothes, K. 1964. The role of kinetin in plant regulation. Régulateurs naturels de la croissance végétale. C.N.R.S., Paris, p. 131–134.

11. Letham, D. S. 1968. A new cytokinin bioassay and the naturally occurring cytokinin complex, p. 19–32. *In* F. Wightman and G. Setterfield [eds.] Biochemistry and physiology of plant growth substances. Runge Press, Ottawa.

12. Hall, R. H., L. Csonka, H. David and B. McLennan. 1967. Cytokinins in the soluble RNA of plant tissues. Science **156**:69–71.

13. Vold, B. S. and P. S. Sypherd. 1968. Modification in transfer RNA during the differentiation of wheat seedlings. Proc. Nat. Acad. Sci. U.S. **59**:453–458.

14. Fox, J. E. and C. M. Chen. 1967. Characterization of labeled RNA from tissue grown on ^{14}C containing cytokinins. J. Biol. Chem. **242**:4490–4494.

15. Chen, C. M. and R. H. Hall. 1969. Biosynthesis of N^6-(Δ^2-isopentenyl) adenosine in the transfer ribonucleic acid of cultured tobacco pith tissue. Phytochemistry **8**:1687–1695.

16. Kende, H. and J. E. Tavares. 1968. On the significance of cytokinin incorporation into RNA. Plant Physiol. **43**:1244–1248.

17. Brandes, H. and H. Kende. 1968. Studies on cytokinin-controlled bud formation in moss protonemata. Plant Physiol. **43**:827–837.

SIX

Abscisic Acid, Dormancy and Germination

To be able to survive in a world of changing environment and constant challenge, the plant must be able to stop as well as to start its various physiological processes. Not only must buds be capable of growth when the favorable weather of the spring arrives, they must also be capable of ceasing activity and preparing for the unfavorable conditions of an approaching freezing winter or a long dry spell. Preparation for these unfavorable periods must be made well in advance of their arrival if the plant is to survive them, since some of the adaptations involve the formation of new substances (e.g., wax), tissues (e.g., periderm) or organs (e.g., bud scales). This necessitates that the plant develop some system for sensing and anticipating the approach of the unfavorable period.

Let us consider how a deciduous tree of the north temperate zone might respond to the challenge of an approaching winter. An examination of a dormant winter twig reveals that the tender terminal growing point is protected from the outside world in several ways (Fig. 6-1). Surrounding it is usually a tuft of cottony insulating material, composed of hairlike organs modified from leaf primordia in the bud. Instead of the usual

135

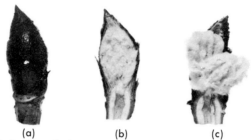

(a)   (b)   (c)

Fig. 6-1   Winter buds of *Aesculus hippocastanum* (horse chestnut) showing the protective devices to withstand the winter conditions. (a) External view—sticky outer scales. (b) Longitudinal section—overlapping bud scales encase the highly compacted young foliage leaves enveloped in cottony down in the center. (c) Two of the young palmate leaves have been spread out to show the thick protective cottony covering.

expanded leaves produced at nodes, there are small, thickened bud scales. The internodes between these bud scales are not at all extended; thus the bud scales overlap in shingle fashion and are closely appressed, forming a tightly fitting protective cap surrounding the insulated growing point. In addition, the scales are covered by a protective varnishlike material that effectively waterproofs the bud and furnishes additional insulative protection. Below the terminal bud, all the leaves have neatly abscised and the leaf scar marking the point of their previous attachment to the stem is well protected by a corky layer that effectively seals the broken vascular bundles and the cells around them. Finally, deep within the trunk, the cambium has become dormant and the parenchyma cells of the xylem and phloem have developed very high osmotic concentrations and have undergone other chemical changes that will make it possible for them to withstand freezing injury.

In the temperate zones, where the approach of the winter is presaged by a progressive diminution of the day length, preparation for the dormant state is accomplished by a sensing of the photoperiod, usually by the mature leaf. Once the photoperiod has declined to less than some critical length (or, more exactly, the length of the dark period has increased to just above some critical length), the metabolism of the leaf undergoes a drastic change. It now starts to make substances that it did not make before. These substances have the general effect of inhibiting or shutting down completely certain crucial components of the synthetic machinery of the leaf. Some of these inhibitory substances are mobile and, moving out into the stem and the growing point, begin to shut down the normal vegetative machinery and prepare the plant for a period of dormancy. By the time the first frost arrives, the plant has adopted its winter, dormant habit and, unless the temperature stresses are too severe, is able to survive until the following spring.

Once spring comes, there are further tactical problems to solve. If, in fact, inhibitory substances have shut down the synthetic machinery of the cells of the growing point, how is one to start them up again? This can be accomplished either by gradually destroying the inhibitor (a kind of chemical hourglass mechanism, in which the rate of disappearance of the inhibitor is nicely linked to the length of the winter frost period) or by producing a substance that competes metabolically with the inhibitor, thus overcoming its action by sheer quantitative preponderance. The plant appears to use both of these techniques.

Discovery of abscisic acid

In the birch tree (*Betula*), Wareing and his colleagues in Aberystwyth, Wales found that the leaves produce ever-increasing quantities of an inhibitory substance as the season progresses into the autumn. This substance could be detected either by its inhibition of auxin-induced growth in various grass coleoptile segments, or by its inhibitory action on the germination of *Betula* seeds. Since artificial shortening of the photoperiod to which the tree was exposed produced the same results, the onset of inhibitor production could reasonably be attributed to photoperiod. Progressive purification of the inhibitor was accomplished by subjecting the various fractions to bioassay and further subdividing the active fractions. Finally, a pure crystalline material was isolated and given the trivial name *dormin*, to indicate its dormancy-inducing activity. Shortly thereafter, a chemical synthesis for this compound was worked out which also independently proved its structure and stereochemistry (Fig. 6-2). The availability of the synthetic material also afforded an independent assay for dormin, based entirely on a physical characteristic, the abnormally high optical rotatory dispersion of the dormin molecule.

Entirely independently, Addicott and his colleagues in Davis, California were investigating the abscission of various organs in plants that were either naturally senescent or were induced to senesce and abscise by chemical or surgical treatment. One of their test objects was the maturing cotton boll, which they found produced a diffusible substance that both promoted abscission and greatly inhibited auxin-induced growth of grass coleoptile cylinders. Successive fractionations and bioassay of each fraction led to the isolation in pure form of an active material. It was named *abscisin II*.

Fig. 6-2 The structural formula of abscisic acid.

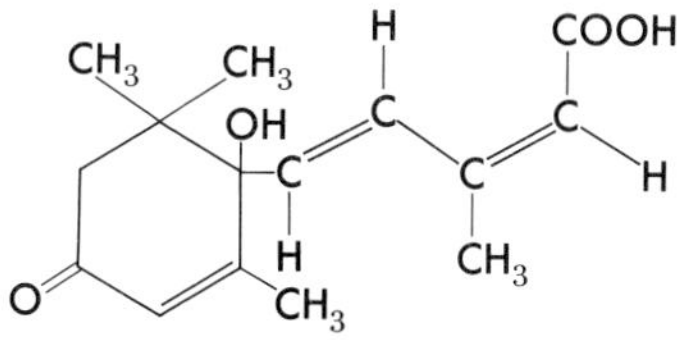

(The first such material investigated, from another source, had been named *abscisin I*. The nature of abscisin I is still obscure.) Abscisin II, isolated only days before Wareing's successful isolation of dormin, proved to be identical with dormin. By mutual consent, this substance is now called *abscisic acid* (ABA). Abscisic acid has now been found in many other species, and there are indications that it is ubiquitous in plants.

Actions of abscisic acid

When abscisic acid is applied to an actively growing twig of *Betula*, or other woody plant, it induces the usual symptoms of approaching

Fig. 6-3   An application of abscisic acid to the apex of a black currant plant has caused the formation of a terminal resting bud. [From El-Antably, Wareing and Hillman (2).]

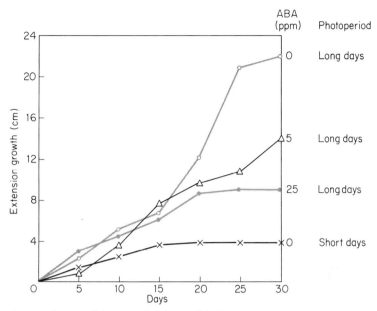

Fig. 6-4 Applications of abscisic acid to shoots of black currant cause a reduction in the extension growth similar to that brought about by placing the plant in short days. [From El-Antably, Wareing and Hillman (2).]

dormancy, i.e., shortened internodes, production of small scalelike leaves instead of expanded foliage leaves, diminished mitotic activity in the apical meristem and abscission of some of the subjacent leaves (1, 2) (Figs. 6-3 and 6-4). It also prevents the bursting of dormant buds (Fig. 6-5) op-

Fig. 6-5 Abscisic acid treatments have delayed bud break in white ash cuttings. Left to right: control, 0.4 ppm, 2.0 ppm, 10.0 ppm abscisic acid after 22 days of treatment. (From Little and Eidt. 1968. Nature **220**:498–499.)

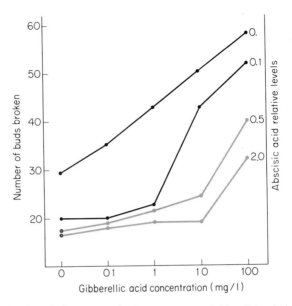

Fig. 6-6   Bud break results from an interplay between promoter (gibberellin) and inhibitor (abscisic acid) levels in the buds. Here the effect of applied gibberellic acid and abscisic acid on bud break in isolated segments of birch stem is examined. [From Eagles and Wareing (1).]

posing the promotive effect of gibberellin (Fig. 6-6) and mimics the effect of decreasing day length with regard to the formation of storage tubers in potato. Thus, it may be reasonably implicated as the active material in the plant's response to shortening photoperiods. Yet this is not the limit of the activity of abscisic acid. When it is applied to seeds capable of germinating, it may act to impose a dormancy upon them (Fig. 6-7). This

Fig. 6-7   Seed germination of tall fescue is inhibited by abscisic acid. (From Sumner and Lyon. 1967. Planta **75:**28–32.)

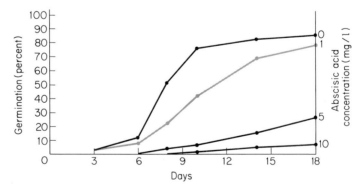

inhibition of germination may be reversed either by extensive leaching of the seeds, or by the application of some material, such as gibberellin or kinetin, which acts in the opposite direction, promoting the germination of many seeds.

We have noted previously that some of the effects of phytochrome-controlled perception of the length of the photoperiod can be understood in terms of the control of production of gibberellic acid. In long-day plants, it will be recalled, GA can in fact substitute for long-day treatments in inducing bolting and flowering of many rosette-type plants. It is thus of great interest to note that ABA, which, like GA, is an isoprenoid substance, is able not only to counteract the effect of gibberellin on flowering in long-day plants, but can also cause flowering in a few short-day plants under noninductive cycles. This indicates that either abscisic acid or some related compound may be at least a part of the flowering hormone in short-day plants.

Several actions of abscisic acid can be counteracted by gibberellins; thus applied GA$_3$ overcame the effect of abscisic acid in the inhibition of elongation of genetically tall corn leaf sections, the germination of numerous seeds (Fig. 6-8) and the sprouting of potato buds. From these opposing activities it was suggested that abscisic acid might specifically antagonize gibberellin in the plant, or might in some cases act as an inhibitor of

Fig. 6-8   The germination of seeds of *Fraxinus ornus* is inhibited by applications of abscisic acid, and this inhibition is counteracted by gibberellic acid. (From data of Sondheimer and Galson. 1966. Plant Physiol. **41**:1397–1398.)

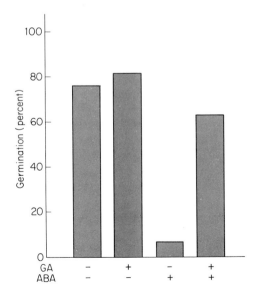

gibberellin biosynthesis (3). Further investigations have shown, however, that the action of abscisic acid is also opposed by cytokinins, and that the effect of the promoter-inhibitor combination may vary not only from species to species but also in the control of growth of different organs of the plants (4). In some cases the interactions become complex. For example, Grand Rapids lettuce seed requires light for germination, and gibberellin can substitute for light. The promotion by gibberellin can be counteracted by abscisic acid, but the inhibitory effect of the latter cannot in turn be overcome by adding increasing concentrations of gibberellin. This proba-bly indicates that the two growth substances do not act competitively at the same metabolic site. The inhibition induced by ABA can, however, be overcome by the addition of kinetin (Fig. 6-9) (5). It appears that the addition of kinetin antagonizes the ABA-induced inhibition and permits gibberellin to promote germination, which then proceeds.

Like gibberellic acid, abscisic acid is composed of isoprenoid units, and can be derived metabolically from mevalonate. Thus, not only do these

Fig. 6-9   Gibberellin promotes the germination of lettuce seed in the dark (B), and the addition of kinetin alone (0.05 mM) has very little effect (A). In the presence of abscisic acid (0.04 mM) (E) the effect of gibberellin is completely removed, and high gibberellin concentrations do not overcome the effect. The addition of kinetin [(C) 0.05 mM; (D) 0.5 mM] in the presence of abscisic acid overcomes the inhibition due to abscisic acid, again allowing the gibberellin to increase germination. [From Khan (5).]

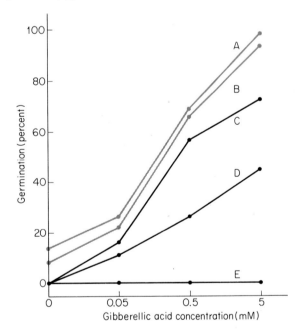

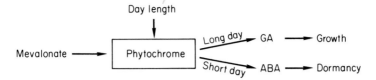

Fig. 6-10   Both abscisic acid and gibberellin are isoprenoid in structure, deriving from mevalonate. Abscisic acid is frequently found under short-day conditions and gibberellin under long days. Phytochrome may possibly be the means by which the synthetic pathway is switched between these two compounds, depending upon the photoperiod.

two compounds interact physiologically, but they are probably formed through a similar metabolic pathway. This may provide a means of control in the plant by a switch from the production of one to the other. Since gibberellic acid is generally formed in long days and abscisic acid in short days, day length may control a metabolic switch, presumably via the phytochrome system (Fig. 6-10) that determines which of these two substances may be synthesized. This would furnish a means by which changes in the environment could lead directly to changes in growth and morphogenesis via control of a single biosynthetic pathway. The fact that applications of abscisic acid can be counteracted by application of other hormones raises the possibility that such effects do occur in the plant, permitting growth control in a precise "stop-go" fashion.

Seed dormancy and germination

Seed dormancy, like the winter dormancy of buds of deciduous trees, has obvious survival value for the plant, for if the seed were to germinate immediately as it matured in the autumn, the tender seedling would almost certainly be nipped by the first frost of the autumn. The incorporation into the seed of a critical quantity of an inhibitor such as abscisic acid will guarantee a lack of activity during the winter season. Many seeds are known in which germination is prevented by inhibitors found either in the seed coat, in persistent organs of the flower or in the tissues of the fruit that surround the seed. Frequently, germination can be induced by the simple technique of exhaustively leaching the seed in running water. Although it appears that many different kinds of inhibitors can be utilized by plants to achieve such dormancy, abscisic acid would seem to be one of the most important. In *Fraxinus americana* (ash), for example, abscisic acid is present in all tissues, but the highest concentrations are in the seed and pericarp of dormant fruits.

The breaking of seed dormancy normally occurs by means of either light or low temperature. The light control of germination, previously discussed, is usually by way of the phytochrome system (Chapter 1). With regard to temperature effects on germination, many seeds require exposure to a period of relatively low temperature (2 to 5°C) for a fairly extended period of 4 to 6 weeks. In horticulture these conditions are frequently supplied artificially, a practice referred to as *stratification.* Both the light and low temperature stimuli are linked to a mechanism that either slowly depletes the inhibitor over the winter or antagonizes it by the production of a stimulatory material, so that the seed will be ready to germinate at the first favorable weather of the spring. During a period of low temperature that breaks dormancy in *Fraxinus americana,* the abscisic acid levels decrease markedly, especially in the seed, which showed a 68% decrease (6). The ABA levels then become as low, in fact, as the concentrations found in seeds of an ash species that does not possess any dormancy. It would thus seem likely that in ash, abscisic acid has a regulatory role in seed dormancy and germination.

The application of gibberellin to light- or cold-requiring seeds can frequently overcome these environmental requirements for germination (Fig. 6-11). It would be reasonable to suppose, therefore, that these environmental changes stimulate the formation of gibberellin. Recent data support this hypothesis. In hazel seeds, chilling, which naturally breaks dormancy, activates the mechanism for gibberellin synthesis (Fig. 6-12) (7). The subsequent synthesis of gibberellin is thought to occur at germination

Fig. 6-11   Gibberellin frequently promotes the germination of seeds, in this instance hazel (*Corylus avellana*). (From Bradbeer and Pinfield. 1967. New Phytol. **66**:515–523.)

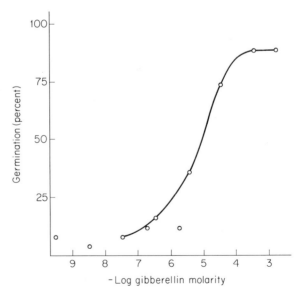

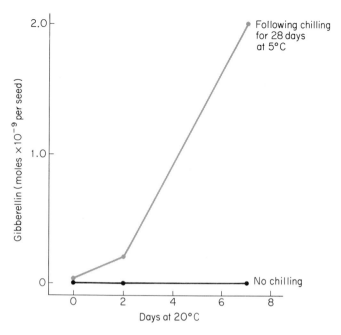

Fig. 6-12   Some seeds require chilling before they will germinate. In hazel (*Corylus avellana*) this triggers the production of gibberellin when the seeds are returned to the germination temperature, which then promotes germination. [From data of Ross and Bradbeer (7).]

temperatures (approximately 20°C) and not at the chilling temperature (approximately 5°C), though in some seeds gibberellin synthesis seems to proceed at a remarkably low temperature. The increase in gibberellin is particularly pronounced in the embryonic axis, suggesting that this might be the principal site of gibberellin synthesis. Illumination by red light, converting phytochrome to the far-red absorbing form (P_{fr}) has also been found to result in an early increase in gibberellin content in lettuce seeds. This is possibly the way in which light induces germination, but the idea has been questioned by the finding that the action of exogenous gibberellin and P_{fr} are synergistic and not additive as would be expected if P_{fr} caused the formation of gibberellin (8). We know that P_{fr} probably also controls membrane permeability and this mechanism might render the applied gibberellins more available at the site of action, providing a basis for the observed synergistic effect. It may be concluded that at least in some cases light or chilling results in increased gibberellin formation. Since many experiments have shown gibberellin to oppose the effects of abscisic acid in growth control, this may reasonably be supposed to represent the mechanism by which the inhibitor is counteracted and germination initiated.

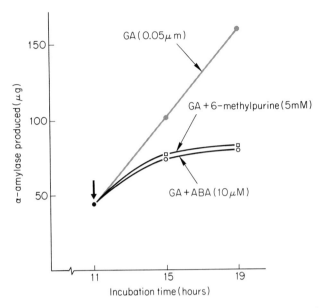

Fig. 6-13   Abscisic acid inhibits the gibberellin-enhanced production of α-amylase with kinetics of inhibition similar to the RNA synthesis inhibitor 6-methylpurine. This indicates that abscisic acid may act by inhibiting RNA synthesis. The inhibitors were added at 11 hours (arrowed). [From Chrispeels and Varner (10).]

Mode of action of abscisic acid

The metabolic site of action of abscisic acid has not yet been elucidated, though, as with several other hormones, some role in the control of nucleic acid and protein synthesis is indicated. Abscisic acid inhibits the synthesis of α-amylase in barley grains and is antagonistic to gibberellin in this process (Fig. 6-13) (9). The kinetics of inhibition of enzyme synthesis resemble that produced by the RNA synthesis inhibitors, 8-azaguanine and 6-methylpurine. This suggests that abscisic acid may exert its action by inhibiting the synthesis of enzyme-specific RNA molecules or by preventing their incorporation into an active enzyme-synthesizing unit. Inhibition of RNA synthesis has indeed been found in *Taraxacum officinale* (dandelion) leaves (10), whereas in *Lemna* (duckweed) abscisic acid inhibits, and cytokinin promotes, DNA synthesis (11). With chromatin isolated from *Raphanus sativus* (radish) hypocotyls, abscisic acid caused about a 28% drop in the amount of uridine monophosphate incorporated into RNA, pointing to an inhibition at the level of RNA transcription (12). This effect was produced only when the abscisic acid was added prior to chromatin extraction, indicating that some binding of the abscisic acid to a cytoplasmic factor may be a prerequisite for action.

Interaction between abscisic acid and promotive hormones could occur at many different points between the initial site of hormone action on nucleic acid metabolism and the ultimate effects on protein synthesis or growth. Thus it is not surprising that many unspecific types of opposing hormonal actions should have been recorded.

Summary

Abscisic acid was originally discovered as a dormancy-inducer in birch and as an abscission-accelerator in cotton. It has a wide range of effects, including induction of dormancy in buds and seeds, promotion of flowering in some short-day plants and diminution of the action of the other promotive hormones. It serves an important function in the plant's life cycle by enabling the plant to shut down activities and survive adverse conditions in a dormant state. The release from dormancy occurs via a drop in inhibitor levels or a rise in promotive hormones that counteract the inhibitor. The promotive hormone is generally gibberellin and its formation is induced either by low temperatures or red light via the phytochrome system. Low temperatures may also lead to decreased inhibitor levels. Abscisic acid appears to act by preventing RNA and protein synthesis.

REFERENCES

General

Addicott, F. T. and J. L. Lyon. 1969. Physiology of abscisic acid and related substances. Ann. Rev. Plant Physiol. **20**:139–164.

Amen, R. D. 1968. A model of seed dormancy. Bot. Rev. **34**:1–31.

1. Eagles, C. F. and P. F. Wareing. 1964. The role of growth substances in the regulation of bud dormancy. Physiol. Plant. **17**:697–709.

2. El-Antably, H. M. M., P. F. Wareing and J. Hillman. 1967. Some physiological responses to D, L-abscisin (Dormin). Planta **73**:74–90.

3. Thomas, T. H., P. F. Wareing, P. M. Robinson. 1965. Action of the sycamore "Dormin" as a gibberellin antagonist. Nature **205**:1270–1272.

4. Aspinall, D., L. G. Paleg, F. T. Addicott. 1967. Abscisin II and some hormone-regulated plant responses. Aust. J. Biol. Sci. **20**:869–882.

5. Khan, A. A. 1968. Inhibition of gibberellic acid-induced germination by abscisic acid and reversal by cytokinins. Plant Physiol. **43**:1463–1465.

6. Sondheimer, E., D. S. Tzou and E. C. Galson. 1968. Abscisic acids levels and seed dormancy. Plant Physiol. **43:**1443–1447.

7. Ross, T. D. and J. W. Bradbeer. 1968. Concentrations of gibberellin in chilled hazel seeds. Nature **220:**85–86.

8. Bewley, J. D., M. Black and M. Negbi. 1967. Immediate action of phytochrome in light-stimulated lettuce seeds. Nature **215:**648–649.

9. Chrispeels, M. J. and J. E. Varner. 1967. Hormonal control of enzyme synthesis: On the mode of action of gibberellic acid and abscisin in aleurone layers of barley. Plant Physiol. **42:**1008–1016.

10. Wareing, P. F., J. Good and J. Manuel. 1968. Some possible physiological roles of abscisic acid, p. 1561–1579. *In* F. Wightman and G. Setterfield [eds.] Biochemistry and physiology of plant growth substances. Runge Press, Ottawa.

11. van Overbeek, J., J. E. Loeffler and M. I. R. Mason. 1967. Dormin (Abscisin II), inhibitor of plant DNA synthesis? Science **156:**1497–1499.

12. Pearson, J. A. and P. F. Wareing. 1969. Effect of abscisic acid on activity of chromatin. Nature **221:**672–673.

SEVEN

Reactions to

Injury

Throughout its life cycle, the plant must respond to challenges from the environment and to changes in the seasons. On some occasions, the plant may be called upon to react to such unusual stresses as physical injury or invasion by pathogens, which threaten the integrity of the plant body. The plant's reactions in these situations are normally such to protect its tissues, to prevent water loss and to set up both physical and chemical barriers to continued progress of the pathogens. The nature of the reaction to the pathogen determines whether the plant is susceptible or resistant to the disease caused by the organism.

Mechanical wounding generally causes either a localized burst of cell division activity or a change in the growth pattern so as to cause the new cells to cover the wound. Alternatively, new specialized scar tissues may be formed after mechanical or chemical injury. A simple example of this is the cork that is caused to differentiate over many exposed wound surfaces, like the cut surface of a potato tuber. The corky layer prevents desiccation of the tuber tissues and is also impervious to fungal attack.

Chemical responses of plant tissue
to wounding or infection

It is now clear that numerous chemical changes are induced by wounding many plant tissues. These changes may, in turn, cause cellular proliferation. In fleshy storage tissues, which have been most extensively investigated, the previously quiescent cells commence extensive synthesis of messenger RNA, resulting in an increase in both the polysome content (Fig. 7-1) and the incorporation of amino acids into protein (Fig. 7-2) (1, 2). These processes are inhibited by the usual RNA and protein synthesis inhibitors (Fig. 7-3), supporting the hypothesis that the response to injury involves a derepression of gene activity.

In sweet potato roots, attack by pathogens or wounding by cutting results in the production within and near the injured cells of phenolic acids [chlorogenic acid, isochlorogenic acid and caffeic acid (Fig. 7-4)] that were not present previously except in the outer cells of the cortex and the cells around the vascular bundles. The amount produced by cut tissues is less than that found in infected tissue. The synthesis of these phenolic acids is mediated by the enzyme phenylalanine ammonia-lyase and, to a lesser extent, tyrosine ammonia-lyase, which deaminate the named amino acids to cinnamic and p-coumaric acid, respectively (Fig. 7-4). Both enzymes are virtually absent from healthy tissue but are formed rapidly after infection or cutting. Their presence is correlated with phenolic acid production (Fig. 7-5) (3).

The enzyme peroxidase has also been found to increase following fungal and bacterial infection of many plants (4). Although peroxidases exist in healthy tissue, the increase in peroxidase activity after injury is largely

Fig. 7-1   The polyribosome content of carrot root discs increases following cutting, indicating an increase in protein formation. [From Leaver and Key (2).]

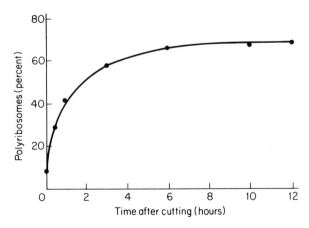

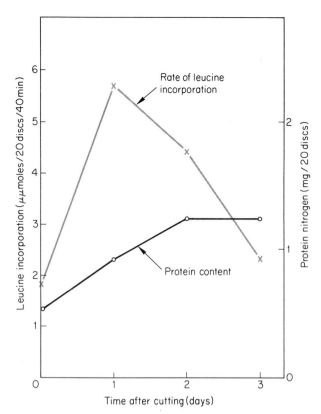

Fig. 7-2 Accompanying the increase in polyribosomes there is an increase in both the rate of amino acid incorporation into protein and the protein content of the tissue. These measurements were made on the microsomal fraction from beet discs. [From Ellis and MacDonald (1).]

Fig. 7-3 The incorporation of adenosine-^{3}H into RNA of polyribosomes (P) with an accompanying decrease in monoribosome (M) content following cutting of carrot root discs is depicted in (a). The mono- and polyribosomes have been separated by a sucrose density gradient. The incorporation of the adenosine and the formation of polyribosomes are both prevented when the excised carrot root discs are incubated in the presence of Actinomycin D (b). [From Leaver and Key (2).]

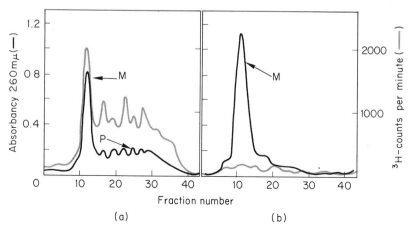

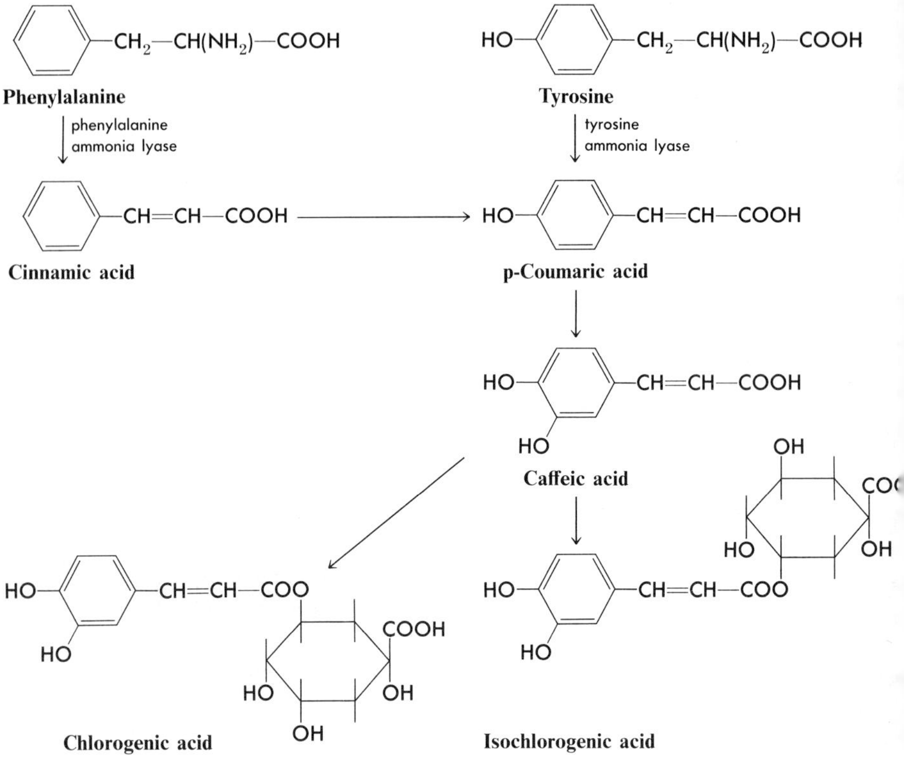

Fig. 7-4 Synthesis pathways of phenolic acids, which develop in certain plant tissues (e.g., sweet potato roots) in response to injury or infection.

caused by the formation of new molecular forms, isozymes, of the enzyme (Fig. 7-6). Inhibitor studies of this phenomenon have indicated that de novo enzyme formation, directed by DNA-dependent RNA synthesis, is involved. The degree to which different varieties respond to fungal infection suggests that the production of peroxidase is a prime factor in disease resistance (Fig. 7-7).

The induction of peroxidase can be brought about not only by a pathogen, but also by nonpathogenic fungi. This appears to be a way in which plants can acquire a type of immunity, since, if an inoculation of a nonpathogenic fungus is followed by a pathogenic-fungal invasion, the enzyme changes induced by the initial inoculation render the tissue at least partially resistant to further fungal attack (Fig. 7-8).

In contrast to antibodies, the immune factors of animals which are carried throughout the blood stream, the defense reaction of plants seems to be restricted to a few cells immediately adjacent to the infection site.

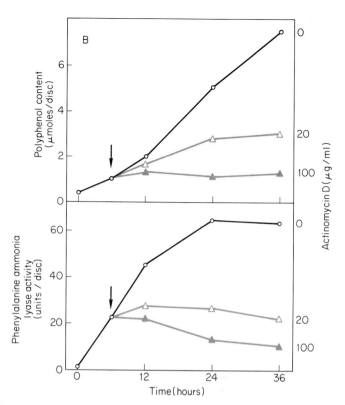

Fig. 7-5 The development of polyphenols in sweet potato in response to injury depends upon the promotion of protein synthesis. Their appearance is correlated with an increase in phenylalanine ammonia lyase activity, and both are inhibited by actinomycin D (applied at 6 hours—arrowed). [From Uritani, et al. (3).]

Fig. 7-6 Bacterial or fungal attack of plant tissues can cause a change in the peroxidase isozyme pattern. The figure shows changes in electrophoretic patterns of peroxidase isozymes from tobacco leaf extracts injected with heat-killed cells of *Pseudomonas tabaci* (b, d) compared with leaves injected with water (a, c). The extraction of (a) and (b) was made immediately after injection; the extraction of (c) and (d) was carried out on the fourth day after injection. (From Lovrekovich, Lovrekovich and Stahmann. 1968. Phytopath. **58**:193-198.)

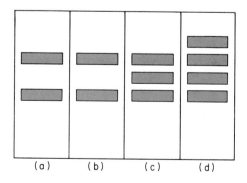

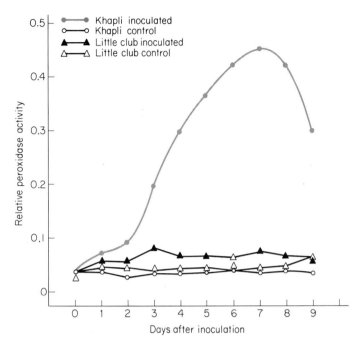

Days after inoculation

Fig. 7-7   The induction of peroxidase by disease infection appears to contribute to conferring resistance to the disease. When the peroxidase activity in wheat leaves is followed after inoculation with *Puccinia graminis* (wheat rust), little change is seen in a susceptible variety (Little Club), whereas a resistant variety (Khapli) shows a considerable increase. (From Macko, Woodbury and Stahmann. 1968. Phytopath. **58**:1250–1254.)

Fig. 7-8   Resistance in susceptible sweet potato (Jersey Orange) to a pathogenic isolate of *Ceratocystis fimbriata* can be induced by inoculation with a nonpathogenic isolate. The upper row shows cross sections of noninoculated root tissue and sections inoculated with the nonpathogenic isolate on the upper and lower surfaces. The bottom row is the same, except that the sections were challenged with the pathogenic isolate 2 days later. Where no immunity was induced, as in the noninoculated susceptible tissue, the entire piece developed black rot (lower left). [From Stahmann (4).]

Induction of immunity by nonpathogen

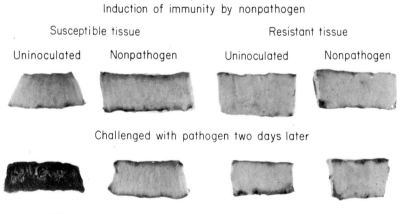

Susceptible tissue   Resistant tissue

Uninoculated   Nonpathogen   Uninoculated   Nonpathogen

Challenged with pathogen two days later

Furthermore, the defense mechanism is not as specific as the antibody response of animals and often results in the destruction of host cells along with those of the parasite. This is relatively unimportant in plants, however, as plants can usually easily replace a tissue area or an organ sacrificed to overcome an infectious agent.

The chemical nature of the
wounding stimulus

The control of the wounding reaction is relatively obscure. If freshly cut plant tissue is immediately rinsed exhaustively with water, very few cell divisions will occur in the cells adjacent to the wound. However, if the wound area is smeared with finely ground tissue of the same species, cell division takes place. This result has led to the postulation of the existence of substances called "wound hormones" that are produced by the injured tissues and are required for cell division of the cells bordering the wound. A compound called traumatic acid (1,10-decene dicarboxylic acid) has been extracted from bean pods. It can cause wound cork formation in washed discs of potato tuber and the formation of intumescences in bean pod placentae. Because it does not produce effects on most other plant tissues, its role has been questioned. It is possible that other hormones such as cytokinins may be more importantly involved in the wound reaction. Additional data on the hormonal control of the wound response are required.

The chemical responses of plant tissues to fungal infection result not only from the physical wounding caused by progress of the mycelium through the tissue but also from the secretion of various chemicals by the pathogen. Auxin and extracts from the fungal tissue have been found to be without effect in inducing immunity to fungi in sweet potato root tissue, but application of the enzymes ribonuclease, acid phosphatase and papain do have such an effect. It is possible, therefore, that the breakdown of macromolecules by ribonucleases and proteases may release compounds that diffuse into adjoining cells and induce the development and expression of immunity.

One plant hormone that has been shown to be involved in the plant's response to infection is ethylene. It was noticed that noninoculated pieces of sweet potato root tissue showed increasing levels of peroxidase if they were stored in the same container as inoculated tissue. Treatment of the tissue with ethylene also induces peroxidase, and ethylene has been detected as a product of sweet potato tissue that has either been injured (5) or infected by black rot fungus (*Ceratocystis*) (6). It thus seems that the ethylene may be one of the initial stimuli which, moving from primary areas of infection into surrounding cells, induce peroxidase formation and increased resistance against certain fungi (4).

The control of peroxidase activity by hormones can be conveniently studied in aseptic cultures of the stem pith of various plants. This tissue is a favorable object for study because it consists initially of homogeneous, nondividing parenchyma cells that can be stimulated to divide and differentiate under appropriate conditions. The young pith cells of the stems of geranium and tobacco lack peroxidase activity completely. In the geranium plant, pith cells never form peroxidase *in situ,* but do so within several hours after excision (7). The nature of this control is not known, but injury appears to derepress the genes controlling peroxidase formation. There seems to be but a single geranium peroxidase component when the enzyme is subjected to electrophoresis; its formation is inhibited by auxin and promoted by kinetin. These two substances interact in the control of peroxidase activity as they do in the control of growth. In tobacco pith, peroxidase appears progressively as the cells age and there is a strict inverse correlation between peroxidase activity and growth of the explant in culture. All of the activity can be resolved into two isozymes (A_1 and A_2) migrating to the anode at pH 9.0 (8). When such tissue is excised and placed in a basal culture medium, it immediately starts to form several new isozymes, including two, C_1 and C_2, which migrate toward the cathode (Fig. 7-9). Auxin represses and kinetin promotes the synthesis of these isozymes (Fig. 7-10). After about 120 hours, still another isozyme (C_3) can be detected, but only if the medium contains auxin. In view of (a) the delay in this induction effect, (b) the known induction of ethylene synthesis by auxin and (c) the induction of peroxidase activity by ethylene, it is possible that C_3's appearance is elicited by ethylene. Thus, it appears that

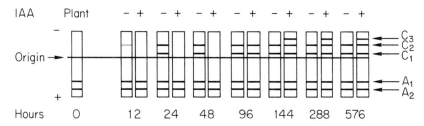

Fig. 7-9   The effect of the auxin indoleacetic acid (IAA) on the electrophoretic isoperoxidase pattern of tobacco stem pith aseptically cultured on modified White's medium. The original pith had two anodic isozymes; within 12 hours in culture, new cathodic isozymes develop, which are repressed by IAA. By 96 hours of culture, this repression has disappeared, and beyond this point a unique, rapidly moving isozyme appears only in cultures containing IAA. [From Galston, Lavee and Siegel (8).]

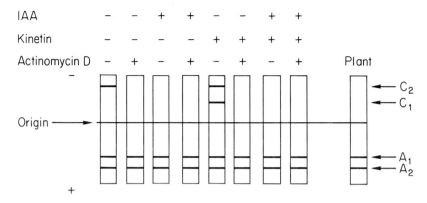

Fig. 7-10 The effect of IAA, kinetin and actinomycin D on isoperoxidase patterns of cultured tobacco pith 24 hours after excision of the tissue. Kinetin alone promotes enzyme formation, while IAA and actinomycin D inhibit the appearance of the new activity. [From Galston, Lavee and Siegel (8).]

at least three different plant hormones, auxin, cytokinin and ethylene, can control the level of peroxidase in tissue that is responding to wounding by cellular activity in aseptic culture. In addition, we should recall that the abnormally high peroxidase levels in genetic dwarfs can be reduced by gibberellin application (Chapter 4).

Response to invasion by pathogens—

Phytoalexins

The growing plant possesses several kinds of barriers to invasion by pathogenic organisms. Such barriers may be either physical (the cuticle covering the epidermis) or chemical (the presence of endogenous fungitoxic compounds, such as phenols). In some cases, however, plants are able to respond defensively to the invasion by a pathogen through the production of new fungitoxic substances that are formed *only* when the host cells come into contact with the parasite. Such compounds have been termed *phytoalexins* (from the Greek *phyton*—plant—and *alexin*—a warding-off compound).

Phytoalexins are formed only by living host cells challenged by invasion. They are nonspecific in fungal toxicity, though the relative sensitivity of the fungi may vary. The reaction is confined to the tissue colonized by the fungus and its immediate neighborhood. The fungitoxic properties of the phytoalexin prevent further growth of the fungus and thus render the cells resistant to the disease.

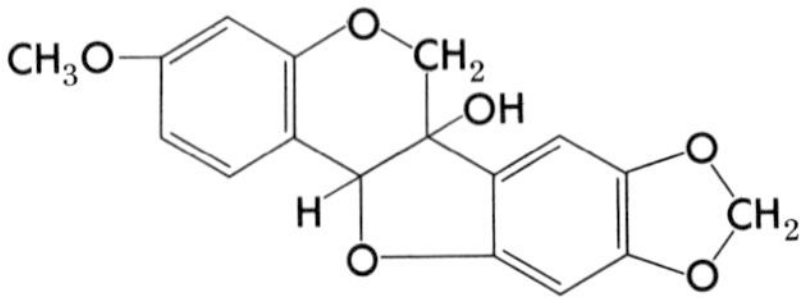

Fig. 7-11   The structure of the phytoalexin pisatin.

That phytoalexins are a product of the host cells is indicated by the fact that they can sometimes also be induced to appear by applications of a range of chemical solutions to the tissue. Another line of evidence supporting this view is that phytoalexins display host and not pathogen specificity. The phytoalexin can be extracted from a wound on a resistant plant and applied to a sensitive plant, where it confers resistance. Several compounds of this type have now been isolated from a wide range of plants and appear to differ greatly in chemical structure. As one example of such a compound we can take *pisatin* (Fig. 7-11), which was originally found following inoculation of endocarp tissue of detached pea pods with the nonpathogenic fungus *Monilinia fructicola.*

The question of the susceptibility or resistance of specific plants or varieties to specific pathogens is dependent upon two factors: the rate and extent of formation of the phytoalexin by the host cells and the sensitivity of the pathogen to the phytoalexin (Tables 7-1 and 7-2*) (1). In general, fungi that are pathogenic are relatively insensitive to the fungitoxin, while fungi that are nonpathogenic are highly sensitive (Fig. 7-12). The levels of both production and sensitivity vary, but the situation required for a resistant reaction is one in which the infected host cells produce a phytoalexin at a concentration above the threshold that inhibits the fungus. In some cases a pathogenic fungus may have the ability to degrade the phytoalexin as it is produced. For example, alfalfa phytoalexin is degraded by the pathogen *Stemphylium botryosum* but not by the nonpathogen *Helminthosporium turcicum.*

Production and action of phytoalexins

The pathway of phytoalexin production depends on the exact phytoalexin under consideration. Pisatin, for example, requires little change in the metabolism of the tissue, since closely related isoflavonoids of similar structure are widespread in *Leguminosae.* Induced pea pods have been found to incorporate radioactivity from phenylalanine, cinnamic

*Both tables are from Cruickshank and Perrin. 1965. Aust. J. Biol. Sci. **18:**829–835.

TABLE 7-1

CONCENTRATION OF PISATIN FORMED BY PEA CULTIVARS INOCULATED WITH
CANBERRA STRAIN OF *Ascochyta pisi*

Pea cultivar	Disease reaction of host tissues	Pisatin concentration (µg/g of endocarp fresh weight)
M. U. 9 ex Tibet	Highly susceptible	309
Little Gem	Highly susceptible	545
Collegian	Highly susceptible	625
Greenfeast	Susceptible	686
Veritable St. Laurent	Susceptible	826
Sans Par Chemin de 40 Jours	Susceptible	821
O.A.C. 181	Semiresistant	1049
Early Blue	Semiresistant	1172
A-100	Semiresistant	1197

Varieties that are more resistant are shown above to form more pisatin.

acid, acetate and methionine into pisatin. In the presence of L-phenyl-alanine-^{14}C the agents that stimulated pisatin production appeared to be acting through the induction of enzyme synthesis in the phenylalanine-to-pisatin pathway (9). This was evidenced by an increase in phenyl-alanine ammonia lyase activity, correlated with an increase in protein synthesis and a greater pisatin accumulation in induced pea tissue than in water-treated tissue. Once formed, pisatin turnover in the pea tissue appears to be minimal or nil even if it is applied to noninduced tissue.

Little is known about the way in which phytoalexins inhibit fungal

TABLE 7-2

CONCENTRATION OF PISATIN FORMED BY THE PEA CULTIVAR LITTLE GEM
FOLLOWING INOCULATION WITH FIVE STRAINS OF *Ascochyta pisi*

Strain of A. pisi	Disease reaction of host tissues	Pisatin concentration (µg/ml of diffusate)
Canberra strain	Susceptible	67.5
Canadian strain I	Semiresistant	88.6
Canadian strain II	Semiresistant	103.6
Canadian strain III	Susceptible	76.8
Canadian strain IV	Susceptible	58.1
Water control		4.9

This shows that less virulent strains evoke more pisatin formation than do virulent strains.

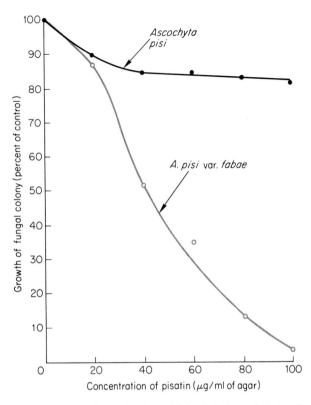

Fig. 7-12   A fungal pathogen of pea, Ascochyta pisi, is relatively unaffected by the pea plant's production of pisatin. Pisatin is, however, extremely effective in preventing the growth of a related variety, A. *pisi var fabae*, with the result that peas are resistant to this variety. (From Cruickshank. 1962. Aust. J. Biol Sci. **15:**147–159.)

growth. Ipomoeamarone, a quinone isolated from sweet potatoes infected with black rot (*Ceratocystis fimbriata*), has been shown to inhibit such diverse metabolic systems as phosphate metabolism, protein synthesis, respiration and growth. The phytoalexin may selectively inhibit the fungus or can be a general growth retardant. Rishitin, a recently isolated phytoalexin from potato, has been found to be inhibitory to growth of oat coleoptiles and wheat leaves, in which it opposes the action of auxin and gibberellin, respectively (Fig. 7-13).

There is a lag phase of 4 to 8 hours (2 to 4 hours after spore germination) between inoculation of pea pods and pisatin production, followed by a rapid rise in pisatin content (Fig. 7-14). The ability of the tissues to produce phytoalexin is also inversely correlated with tissue age, so that older tissues are more susceptible to the invading pathogen. Though the resistant state

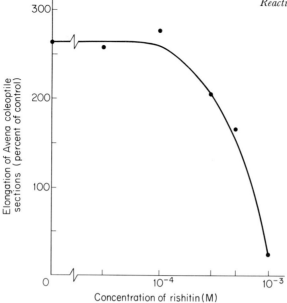

Fig. 7-13   The phytoalexin rishitin counteracts the promotive effect of IAA on the elongation of *Avena* coleoptile sections. (From Tomiyama et al. 1968. Phytopathology **58**:115–116.)

Fig. 7-14   The inoculation of pea pod endocarp with a nonpathogen of pea, *Botrytis allii* causes the induction of pisatin formation following a lag period. (From Cruickshank and Perrin. 1963. Aust. J. Biol. Sci. **16**:111–128.)

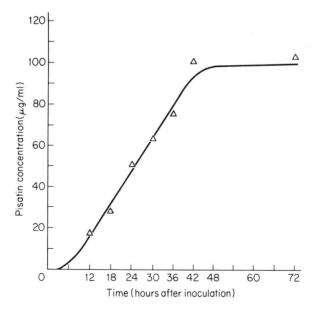

itself is not directly inherited, the capacity to resist fungal attack through phytoalexin production is under genetic control. For example, the resistance of corn to *Helminthosporium turcicum* is under the control of a single gene, which regulates the formation of a phytoalexin in the resistant plant when host and pathogen interact (10).

Summary

    In addition to receiving physical signals from the environment, the plant is also capable of responding to the environmental stresses of wounding or invasion by microorganisms. In both cases, a genetic derepression appears to occur, resulting in enzyme production and tissue proliferation in response to wounding, and production of fungitoxic compounds called *phytoalexins* in response to invasions. The hormonal control of these events is obscure, but when tissue is cut and placed in culture, the appearance of specific peroxidase components is altered by the presence or absence of kinetin and auxin and probably ethylene.

    The formation of phytoalexins by fungal invasion is probably a major mechanism for disease resistance in plants. Nonpathogenic fungi are more sensitive to the phytoalexin than pathogenic species. Resistance on the part of the host plant depends on the inherited capacity to produce adequate amounts of phytoalexins and the speed with which they are produced in response to invasion.

REFERENCES

General

Cruickshank, I. A. M. 1963. Phytoalexins. Ann. Rev. Phytopathology **1**:351–374.

1. Ellis, R. J. and I. R. MacDonald. 1967. Activation of protein synthesis by microsomes from aging beet discs. Plant Physiol. **42**:1297–1302.

2. C. J. Leaver and J. L. Key. 1967. Polyribosome formation and RNA synthesis during aging of carrot root tissue. Proc. Nat. Acad. Sci. U.S. **57**:1338–1344.

3. Uritani, I., T. Asahi, T. Minamikawa, H. Hyodo, K. Oshima and M. Kojima. 1967. The relation of metabolic changes in infected plants to changes in enzymatic activity, p. 342–356. *In* C. J. Mirocha and I. Uritani [eds.] The dynamic role of molecular constituents in plant-parasite interaction. American Phytopathological Society, St. Paul, Minn.

4. Stahmann, M. A. 1967. Influence of host-parasite interactions on proteins, enzymes and resistance. *Ibid.*, p. 357–369.

5. Imaseki, H., I. Uritani and M. A. Stahmann. 1968. Production of ethylene by injured sweet potato root tissue. Plant and Cell Physiol. 9:757–768.

6. Imaseki, H., T. Teranishi and I. Uritani. 1968. Production of ethylene by sweet potato roots infected by black rot fungus. Plant and Cell Physiol. 9:769–781.

7. Lavee, S. and A. W. Galston. 1968. Hormonal control of peroxidase activity in cultured *Pelargonium* pith. Amer. J. Bot. 55:890–893.

8. Galston, A. W., S. Lavee and B. Z. Siegel. 1968. The induction and repression of peroxidase isozymes by 3-indoleacetic acid, p. 455–472. *In* F. Wightman and G. Setterfield [eds.] Biochemistry and physiology of plant growth substances. Runge Press, Ottawa.

9. Hadwiger, L. A. 1967. Changes in phenylalanine metabolism associated with pisatin production. Phytopath 57:1258–1259.

10. Lim, S. M., J. D. Paxton and A. L. Hooker. 1968. Phytoalexin production in corn resistant to *Helminthosporium turcicum*. Phytopath 58:720–721.

EIGHT

Senescence and
Abscission

Leaf abscission is one of the most complex processes occurring in the plant. It may occur progressively as a plant ages, in which case leaves are being continuously replaced by the stem apex, or it may occur completely and all at once as an overwintering mechanism in a deciduous tree. As almost all the known hormones and regulatory mechanisms seem to enter into its control, it seems worthwhile to discuss it in some detail.

Abscission may be considered as related to and part of the much wider phenomenon of senescence. Relatively little attention has been paid to plant senescence in the past, yet the fact that it is under precise physiological control compels closer scrutiny. Consider, for example, the fact that in the wheat fields of Kansas millions of plants senesce and die at approximately the same time, following the formation of seed. What has caused these plants to cease acitivity and die, leaving only the seeds to carry on in the next season? Why do long-lived perennial species such as the century from the loss of genetic material; rather, they appear to result from, or at least to be correlated with, a series of changes in the pattern of protein series of events leading to the death of the plant; this supposition is reinforced by the fact that removal of the flowers or developing fruits frequently significantly delays or entirely prevents senescence. As the developing fruits and seeds require large stores of nutrients, it was at first considered that the rest of the plant declined because of nutrient deficiency resulting from competition among plant organs. This theory was made less likely when it was shown that even the formation of small male flowers in the dioecious spinach plant led to senescence, and that senescence is not usually triggered while the fruit is growing but only when it begins ripening.

Whether in the individual organ or the whole plant, senescence involves declining metabolic rates and decreasing rates of RNA and protein synthesis. We have already mentioned the changes in respiration rate and membrane permeability that accompany fruit ripening. The action of most hormones is to delay senescence, in part, at least, through maintenance of the synthesis of RNA and protein. For example, in fruit tissue such as bean endocarp, senescence is delayed by either auxin or kinetin; in detached leaves kinetin alone is effective. Many studies indicate that senescence in plants is not a simple running-down and fading away process but is rather an active physiological stage of the life cycle, as much controlled by hormones as any that went before. In plants, death of individual cells or tissues may be a normal, controlled and localized event, helping in the production of the final form of the plant. One example is the death of tracheids and vessel cells, to form the efficient but dead hollow cells of the water-conducting system.

From the time of germination onward, the plant passes through a series of developmental changes, each directed through selective control of parts of its genetic apparatus. We have already shown that induction and repression of particular genes may be controlled by plant hormones (e.g., gibberellin), but we still do not understand the master control system that programs the induction and repression of specific genetic activities. Even in mature, fully differentiated cells, the complete genetic complement exists, containing all the information required to form a whole plant. This potential can be elicited when certain mature cells are placed in tissue culture, where they are stimulated to dedifferentiate, to reinitiate cell division and ultimately to give rise to a whole new plant. From this experiment it is evident that cell maturation and senescence do not result from the loss of genetic material; rather, they appear to result from, or at least to be correlated with, a series of changes in the pattern of protein synthesis in the cell. Cellular senescence may thus result, for example, from altered relative activity of different genes, leading to a failure to produce sufficient m-RNA to maintain those functions essential to cellular integrity, while permitting the synthesis of excessive quantities of various degradative enzymes such as nucleases and proteases. Frequently, senescence is also accompanied by an alteration of membrane permeability, leading to leakage from one cellular compartment to another of materials that should really be kept separated. In many cells, there appear to be membrane-bounded sacs, called *lysosomes*, whose dissolution releases into the cytoplasm

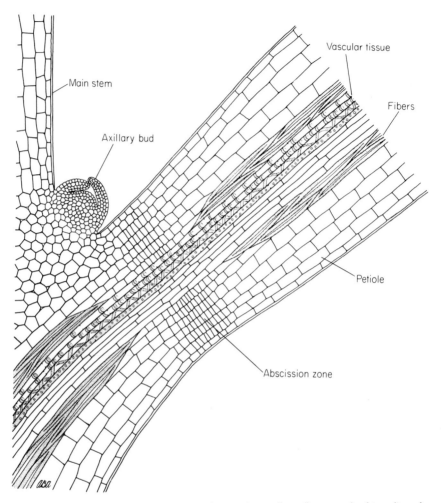

Fig. 8-1   Drawing of the abscission zone showing the smaller cells across the future line of separation. As the tissues age, separation of the cells becomes apparent, followed by abscission of the distal portion of the petiole, leaving intact cells at the abscission interface. (Courtesy of F. T. Addicott, University of California, Davis.)

degradative enzymes which then lead to cellular disruption, senescence and death. We have, unfortunately, little information about the control of lysosome integrity.

Leaf abscission

Leaf abscission results from the formation of specialized cell layers of the abscission zone at, or close to, the base of the petiole (Fig. 8-1). Depending upon the species, this zone may be formed early in the development of the leaf or only after the leaf is fully mature. The parenchyma cells comprising the abscission layer are frequently smaller than the surrounding cells. Even the individual vascular elements may be shorter and fibers may be absent from the bundle in the abscission zone. These anatomical features make this zone an area of weakness.

Leaf fall is initiated by senescence of the leaf, either because of natural aging or, in deciduous trees, by a signal from the environment. This signal is most usually decreasing day length, which triggers changes within the leaf lamina that result in the entire leaf being shed from the tree. Photoperiodic control can be easily seen by the fact that, in the autumn, trees near street lamps frequently retain their leaves longer than other trees. Eventually, of course, declining temperatures insure the abscission of even these leaves.

Prior to leaf fall, numerous changes occur in the abscission zone. Cell divisions frequently occur, forming a layer of brick-shaped cells across the base of the petiole. Active metabolic changes occur in the abscission zone cells (see p. 170) which result in the partial dissolution of the cell wall or the middle lamella. The cells thus become separated; the weight of the leaf snaps the vascular connection and the leaf is shed from the tree. A cork layer forms across the petiole stump, thus protecting the tissues of the tree from microbial invasion and restricting water loss. The xylem vessels become plugged with tyloses, completing the sealing-off process.

Experiments on factors involved in leaf abscission are generally carried out on petiole explants. The leaves are cut off and then either the petiole is excised at the base or a portion of the stem with the petioles is taken (Fig. 8-2). The cut end of the petiole is then treated with a drop of the hormone solution under investigation and the sections are placed in a petri dish with a means of keeping them moist for the duration of the experiment. For ethylene treatments the gas is passed into a sealed container with the explants. Structural or biochemical changes are then followed and after the required period abscission of the distal end of the explant is recorded. In *Phaseolus* (French bean) explants the pulvinus on the distal side of the abscission zone is pushed off by the underlying tissues, but in cotton abscission is checked by the effect of a standard pressure against the end of the petiole. The results are usually recorded as the percentage of the explants that show abscission or the time required for abscission of 50% of the explants.

The control of abscission is still not clearly understood, but it is evident that both auxin and ethylene are prominently involved. During the active life of the leaf, auxin is constantly transported from the lamina through the petiole, but as the leaf ages, auxin production declines (Fig. 8-3). Auxin applied to the lamina will delay abscission by maintaining the metabolic activity of the cells of the petiole. It has been suggested that decreasing supplies of cytokinins from the roots, as winter approaches, may

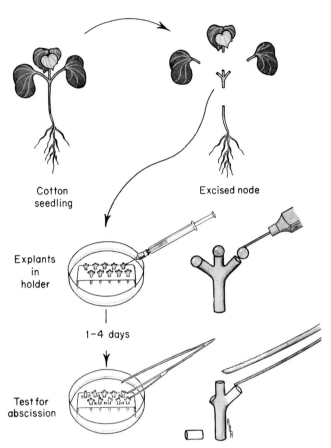

Fig. 8-2 The hormonal control of abscission is studied in explants with the hormone applied to the cut end of the explant. In cotton the petioles are used still attached to the stem and are checked for abscission by using a standard pressure from modified forceps. (From Addicott, et al. Régulateurs naturels de la croissance végétale. C.N.R.S., Paris, 1964.)

also be involved in the onset of senescence in the leaves. This view is supported by the demonstrated ability of cytokinins to lengthen the life of excised leaves, but as yet there are no unambiguous experimental data to support this hypothesis. As the leaf approaches senescence, products from the senescent tissues of the leaf diffuse back down the petiole and cause the onset of senescence in the petiole tissues. There has been some speculation that these "senescence factors" include abscisic acid, but it is generally considered that abscisic acid, despite its name, has little control over abscission.

The aging of tissues is important not only in its control of the production of senescence factors by leaves but also in the response of the abscission zone tissues to hormones. When senescence is retarded in pulvinar cells

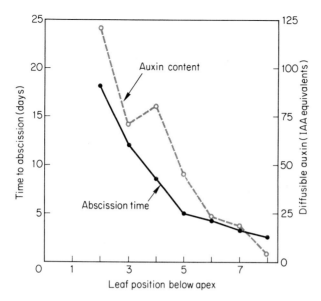

Fig. 8-3 With increasing age, indicated here by the leaf position below the apex, *Coleus* leaves show decline in the time required for the development of abscission upon being debladed, and an associated decline in diffusible auxin content. (From de la Fuente and Leopold. 1968. Plant Physiol. **43**:1496–1502.)

by auxin, abscission is also retarded, and any treatment accelerating senescence (e.g., ethylene) stimulates abscission. Ethylene, however, promotes petiole abscission only if applied after the petiole tissues have aged (Fig. 8-4) (1). Auxin generally inhibits abscission, but can have promotive

Fig. 8-4 When petiole explants are incubated in ethylene, the gas has very little effect on unaged explants. If, however, the explants are aged for 24 hours before ethylene treatment, then ethylene produces a considerable increase in abscission. [From Abeles (5).]

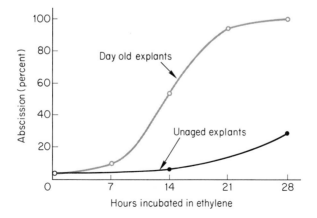

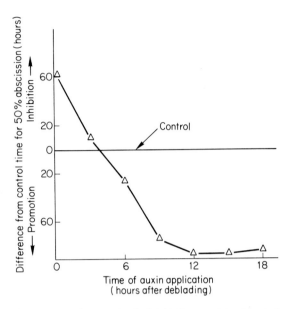

Fig. 8-5   The time of auxin application after leaf deblading can determine whether abscission is inhibited or promoted. Auxin inhibits abscission when applied early following deblading, but once senescence has proceeded auxin promotes abscission. In the untreated control 50% abscission occurred in 136 hours. (From Rubinstein and Leopold. 1963. Plant Physiol. **38**:262–267.)

effects, depending on the age of the tissue to which it is applied (2). When applied to petioles from which the leaf blades have been excised, auxin delays abscission if given during the first few hours after deblading; thereafter it promotes abscission (Fig. 8-5). It appears that as the leaf approaches senescence, the inhibitory effect of auxin is lost and abscission becomes increasingly responsive to the promotive effects of other factors. It has been suggested, therefore, that abscission results in part from an increase in sensitivity to constantly produced ethylene (1). This theory fails, however, to account for the promotion of abscission by auxin under some conditions.

Abscission has been shown to involve an active dissolution of the walls of the cells of the abscission zone, not just a passive break. This results from an increase in the activities of enzymes in cells of the abscission zone, particularly enzymes involved in the breakdown of the cell wall (Fig. 8-6). The separation at the abscission zone has been found to result from an enhanced RNA synthesis and enzyme activity localized in the tissues adjacent to the pending plane of separation (Fig. 8-7) (3). An increase in cellulase activity also occurs in the cortical cells at the pulvinar-petiole tissue interface, but only when senescence takes place in cells distal to the interface (Tables 8-1 and 8-2) (4). Ethylene, which promotes abscission,

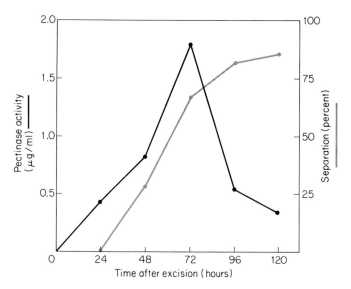

Fig. 8-6   Pectinase activity in explants increases following excision, contributing towards abscission in the explant. (From Morré. 1968. Plant Physiol. **43**:1545–1559.)

also enhances localized RNA and protein synthesis (Fig. 8-8) (5), while inhibitors of RNA and protein synthesis prevent abscission in the presence of ethylene (Fig. 8-9).

Abscission, therefore, results from a complex interplay of hormones, including senescence factors, ethylene and auxin. Auxin produced in young leaves inhibits abscission, and the decline in auxin production as the leaf ages has been implicated in abscission, especially since applied auxin may

TABLE 8-1

THE DEVELOPMENT OF CELLULASE ACTIVITY
ACROSS THE ABSCISSION ZONE

| Segment | *Cellulase activity* | |
	Control	*2,4,5-T*
Pulvinus	19.3	8.3
Abscission zone	46.5	6.6
Petiole	20.1	6.9

When petiolar explants from *Phaseolus* are allowed to age, an increase in cellulase activity can be detected, principally in the abscission zone, much less being present in the tissues on either side of the zone. This increase in cellulase is completely eliminated when tissue senescence is prevented by treating the explants with auxin (2,4,5-T). [From Horton and Osborne (4).]

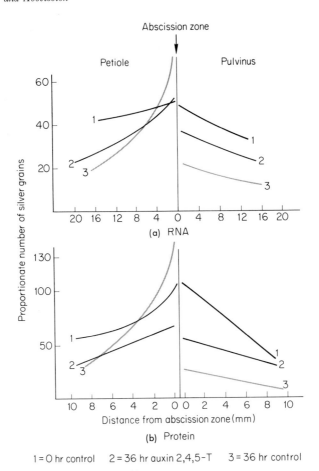

Fig. 8-7 The localized formation of RNA and protein in abscission zones can be seen following the treatment of petiole explants with radioactive precursors and then the use of histoautoradiography to localize the points of incorporation of the precursors into the macromolecules. The figures show the distribution of silver grains for 10 mm on either side of the abscission zone in histoautoradiographs of longitudinal sections of explants exposed to (a) ^{14}C adenine to measure RNA formation, (b) ^{14}C leucine to measure protein formation, at 0 and 36 hours after excision, with the explants either untreated or treated with auxin (2,4,5-T) to retard abscission. [Data from Osborne (3).]

inhibit the process. However, a burst of auxin production has been noted during senescence of detached leaves (Fig. 8-10), possibly due to the death of particular cells (6). The promotion of abscission by auxin under certain conditions may, therefore, result from auxin-stimulated ethylene production, since such an action of auxin can be decreased by the removal of the accumulated ethylene (1). The ethylene then brings about dissolution in abscission zone tissue already aged through the action of senescence

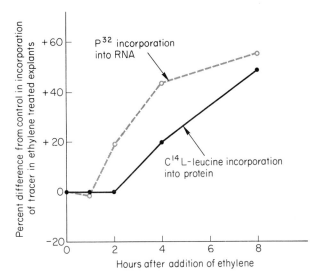

Fig. 8-8   The treatment of petiole explants with ethylene to increase abscission also increases the formation of RNA and protein, as witnessed by the incorporation of precursors. This is interpreted as an enhanced enzyme formation under the influence of ethylene, which leads to the degradation of the cell walls of the abscission zone. [From Abeles (5).]

factors diffusing from the leaves. Alternatively it has been hypothesized that senescence factors themselves are responsible for the rise in the enzyme formation involved in abscission (3). This would eliminate abscisic acid as a senescence factor, since abscisic acid has been shown to inhibit RNA and protein synthesis (Chapter 6).

TABLE 8-2

THE EFFECT OF AUXIN AND ETHYLENE ON
THE DEVELOPMENT OF CELLULASE
ACTIVITY IN THE ABSCISSION ZONE

Treatment	Cellulase activity
Control	38.8
2,4,5-T	8.8
Ethylene	62.6

The cellulase activity in the abscission zone which develops as the explants age (control) is reduced by auxin (2,4,5-T) treatment which delays senescence and is increased by ethylene treatment which accelerates senescence. [From Horton and Osborne (4).]

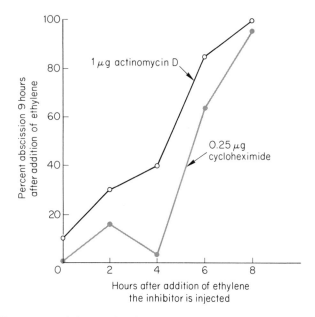

Fig. 8-9 The promotion of abscission by ethylene can be prevented by the addition of actinomycin D or cycloheximide to inhibit RNA and protein synthesis, respectively. If the application of the inhibitors is delayed, the ethylene is able to cause the formation of more and more degradative enzymes so that the inhibitors have progressively less effect. [From Abeles (5).]

Fig. 8-10 Though, in general, auxin levels in leaves decrease as the leaves age, a burst of auxin synthesis has been detected in detached *Avena* leaves just prior to their death. [From Sheldrake and Northcote (6).]

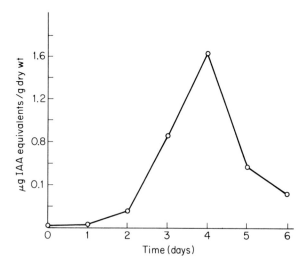

Summary

The senescence and death of entire plants or plant parts are the result of a precisely programmed series of events, frequently under hormonal control. Leaf abscission occurs at a special weak layer of cells at the base of the petiole, usually in response to an environmental signal such as photoperiod. Auxin coming from the leaf blade retards abscission, but once senescence has started, such auxin may promote abscission. This is probably through the formation of ethylene, which in turn stimulates the synthesis of new enzymes, such as cellulase, promoting the dissolution of cell walls in the abscission zone.

REFERENCES

General

Aspects of the biology of ageing. 1967. Symp. Soc. Exp. Biol. 21. Cambridge University Press. 634 p.

Varner, J. E. 1965. Death, p. 867–874. *In* J. Bonner and J. E. Varner [eds.] Plant biochemistry. Academic Press, New York.

Symposium on Leaf Abscission. 1968. Plant Physiology **43**:1471–1586.

1. Abeles, F. B. 1967. Mechanism of action of abscission accelerators. Physiol. Plant. **20**:442–454.

2. Rubinstein, B. and A. C. Leopold. 1963. Analysis of the auxin control of bean leaf abscission. Plant Physiol. **38**:262–267.

3. Osborne, D. J. 1968. Hormonal mechanisms regulating senescence and abscission, p. 815–842. *In* F. Wightman and G. Setterfield [eds.] Biochemistry and physiology of plant growth substances. Runge Press, Ottawa.

4. Horton, R. F. and D. J. Osborne. 1967. Senescence, abscission and cellulase activity in *Phaseolus vulgaris*. Nature **214**:1086–1088.

5. Abeles, F. B. 1968. Role of RNA and protein synthesis in abscission. Plant Physiol. **43**:1577–1586.

6. Sheldrake, A. R. and D. H. Northcote. 1968. Production of auxin by detached leaves. Nature **217**:195.

Index